New York, New York Columbus, Ohio Chicago, Illinois Woodland Hills, California

Illustrators: *Gregory Benton, Ariel Bordeaux, Jim Callahan, Mark Carolan, Wayne Honath, Pat Lewis, Ellen Lindner, Mitch O'Connell, John Pham, Joel Priddy, Brian Ralph, Mark Ricketts, Scott Rolfs, Rob Ullman, Mark Zingarelli*

The McGraw·Hill Companies

Send all inquiries to:

Glencoe/McGraw-Hill
8787 Orion Place
Columbus, OH 43240-4027

ISBN: 978–0–07–877746–2

MHID: 0–07–877746–1

Printed in the United States of America.

1 2 3 4 5 6 7 8 9 10 024 10 09 08 07

TABLE OF CONTENTS

UNIT 5 Africa South of the Sahara

UNIT 6 South Asia

UNIT 7 East Asia and Southeast Asia

UNIT 8 Australia, Oceania, and Antarctica

USING GRAPHIC NOVELS:

POPULAR CULTURE AND SOCIAL STUDIES INTERACT

Graphic novels represent a significant segment of the literary market for adolescents and young adults. These stories may resemble comic books, but on closer inspection, they often address controversial issues using complex story lines. Some graphic novels that are well known to Western audiences are Watchmen, which examines how superheroes live in a society that has turned against them; Maus, which uses anthropomorphic characters to tell the story of a Holocaust survivor; From Hell, which presents one explanation for the actions of the historical serial killer Jack the Ripper; and Road to Perdition, which was made into a popular motion picture.

WHAT ARE GRAPHIC NOVELS?

Graphic novels, as they are known in Western countries, were initially inspired by Japanese *manga* (comics) and *anime* (animation). *Anime* style is most commonly recognizable in its use of large-eyed characters with oversized heads, and it has been increasingly recognized by Western audiences as a distinct art form.

Use of the *manga* genre in Japan is far more widespread than in Western countries and dates back to the early part of the 1900s. Japanese *manga,* rendered in black and white and printed on newsprint, are read by children and adults and include many topics, although science fiction *mechas* (robots) dominate the field. The topics of these works are surprisingly similar to Western young adult fiction. A large portion of the market is *shojo,* comic books designed to appeal to girls. A popular *shojo* character that appears in America is Sailor Moon, a resourceful Japanese schoolgirl. *Shonen manga* is designed primarily for boys and usually consists of action stories. Teachers may recognize elements of *shonen manga* in Japanese game cards collected and traded by many American youth. Many *manga* are published in serial form in books as long as 750 pages. One of the first *manga* marketed for Western consumption was *The Four Immigrants Manga: A Japanese Experience in San Francisco, 1904–1924* (Kiyama, 1999), first published in 1931. It is not in the *anime* style of today's novels, but offers a poignant portrayal of the challenges facing Asian immigrants at the time.

Why Do Graphic Novels Appeal to Students?

Part of the appeal of graphic novels lies in their "underground" (and therefore forbidden) reputation. Another part of the appeal of *manga* and *anime* lies in their sophisticated story lines and the development of complex characters (Izawa, 2002). Unlike American comic books that feature a superhero with fixed and exaggerated attributes, many of these Japanese stories include a subtext of universal themes involving ethical and moral dilemmas. These *gekiga* (literary novels) are ambitious in their scope and intricacy and are becoming more available in English translations. Unlike the broad range of stories available in Japan, however, the stream of *manga* and *anime* reaching Western readers is not so diverse. The bulk of *manga* and *anime* available in America is often skewed toward violent and sexually graphic titles (called *hentai,* or "perverse") that do not reflect the wide range of quality available.

Graphic novels continue to develop and diversify (Frey & Fisher, 2004). Interactive graphic novels presented in serial form are appearing on the Internet. Readers have a number of options when they visit the site each month to view the next installment, such as engaging in role-playing games, creating new characters to interact with those developed by the author, and visiting an extensive catalog for background information. Most of these Web-based graphic novels have decidedly adult content, although users are likely to be Web-savvy adolescents. A unique subset of these graphic novels and *manga* is a style of writing called *fanfiction,* in which readers create and post their own alternative versions of stories featuring their favorite characters (e.g., Chandler-Olcott & Mahar, 2003).

Why Use Graphic Novels in Social Studies?

Graphic novels are amazingly diverse, in terms of both their content and their usefulness. For example, Gorman (2002) notes that graphic novels are exactly what teens are looking for: they are motivating, engaging, challenging, and interesting. Schwartz (2002b, 2004) believes that graphic novels are engaging because they allow teachers to enter the youth culture and students to bring their "out of school" experiences into the classroom. The purpose of such **"multiple literacies"** is to bridge the gap between students' school literacy and the ways in which they use reading and writing outside of school.

Graphic novels have also been used effectively with students with disabilities, students who struggle with reading, and English language learners (e.g., Cary, 2004; Frey & Fisher, 2004; Schwartz, 2002a). One of the theories behind the use of graphic novels for struggling adolescent readers focuses on their effectiveness in presenting complex ideas while reducing reading demands. As a result, all students can thoughtfully discuss the content at hand. As Weiner (2003) noted,

> Graphic novels have found their way into the classroom, as teachers are realizing their usefulness as literacy tools. After a study of graphic novels, researchers concluded that the average graphic novel introduced readers to twice as many words as the average children's book. This realization has reinforced the idea that the comics format is a good way to impart information. (p. 61)

Conclusions

While controversy about graphic novels persists—especially among people who worry that graphic novels will bring the end of traditional book reading—our experiences with adolescents, as well as a number of current research studies, suggest that graphic novels are an important adjunct in our instruction. Graphic novels are viable options for students with disabilities, struggling readers, and English language learners, but they are more powerful than that. Graphic novels are motivating and engaging for all students. They allow us to differentiate our instruction and provide universal access to the curriculum. We hope you'll find the graphic novels in this book useful as you engage your students in the study of history and social studies.

Sincerely,

Douglas Fisher & Nancy Frey

Douglas Fisher, Ph.D.
Professor
San Diego State University

Nancy Frey, Ph.D.
Assistant Professor
San Diego State University

References

Cary, S. (2004). *Going graphic: Comics at work in the multilingual classroom.* Portsmouth, NH: Heinemann.

Chandler-Olcott, K., & Mahar, D. (2003). Adolescents' anime-inspired "fanfictions": An exploration of multiliteracies. *Journal of Adolescent & Adult Literacy,* 46, 556–566.

Fisher, D., & Frey, N. (2004). *Improving adolescent literacy: Strategies at work.* Upper Saddle River, NJ: Merrill Education.

Frey, N., & Fisher, D. (2004). Using graphic novels, anime, and the Internet in an urban high school. *English Journal, 93*(3), 19–25.

Gorman, M. (2002). What teens want: Thirty graphic novels you can't live without. *School Library Journal, 48*(8) 42–47.

Izawa, E. (2004). *What are manga and anime?* Retrieved December 5, 2004, from www.mit.edu:8001/people/rei/Expl.html.

Kiyama, H. Y. (1999). *The four immigrants manga: A Japanese experience in San Francisco, 1904–1924.* Berkeley, CA: Stone Bridge Press.

Schwarz, G. (2002a). Graphic books for diverse needs: Engaging reluctant and curious readers. *ALAN Review, 30*(1), 54–57.

Schwarz, G. E. (2002b). Graphic novels for multiple literacies. *Journal of Adolescent & Adult Literacy, 46,* 262–265.

Schwarz, G. E. (2004). Graphic novels: Multiple cultures and multiple literacies. *Thinking Classroom, 5*(4), 17–24.

Weiner, S. (2003). *The rise of the graphic novel: Faster than a speeding bullet.* New York: Nantier Beall Minoustchine Publishing.

Teaching Strategies for Graphic Novels

As we have noted, graphic novels are an excellent adjunct text. While they cannot and should not replace reading or the core, standards-based textbook, they can be used effectively to build students' background knowledge, to motivate students, to provide a different access route to the content, and to allow students to check and review their work.

Strategies for using graphic novels in the classroom include the following:

1. **Previewing Content.** In advance of the text reading, you can use a graphic novel as a way to activate background information and prior knowledge. For example, you may display a graphic novel on the overhead projector and discuss it with the class. Using a teacher think-aloud, in which you share your thinking about the graphic novel with the class, you might provide students with advance information that they will read later in the book. Alternatively, you may display the graphic novel and invite students, in pairs or groups, to share their thinking with one another. Regardless of the approach, the goal is to activate students' interest and background knowledge prior to the reading.

2. **Narrative Writing.** Ask students to read one of the graphic novels, paying careful attention to the details and imagery used. Then ask each student to write his or her own summary of the story being told in this novel. Graphic novels without much dialogue provide an opportunity for students to create their own dialogue, based on what they know of the content and characters. Not only does this engage students in thinking about the content, but it also provides you with some assessment information. Based on the dialogue that the students write, you'll understand what they already know, what they misunderstand from the story, and what they do not yet know.

3. **Summarizing Information.** A third possible use of graphic novels involves writing summaries. Like oral retellings of readings, written summaries require that students consider the main ideas in a piece of text and use their own words to recap what they know (Frey, Fisher, & Hernandez, 2003). Students can discuss the graphic novel and the text they have read with a small group, and then create their own summaries. Alternatively, students could summarize the text and then create a compare-and-contrast graphic organizer in which they note the differences between their summary of the text and the way that the author/illustrator of the graphic novel summarized the text (e.g., Fisher & Frey, 2004).

4 Reviewing content. In addition to serving as fodder for written summaries, graphic novels can be used for review of content. While there are many reasons to review content—such as preparing for a test—graphic novels are especially useful for providing students with a review of past chapters. You can use a graphic novel from a previous chapter to review the major events in time or place, so that students can situate the new information they are reading in a context.

5 Analysis. Graphic novels often have a thematic strand that illustrates a specific point about the content being studied. This may take the form of irony, humor, or a more direct and formal approach to a historical event. In the analysis approach, students read the graphic novel with the intention of trying to understand the main point the author is trying to convey. This approach is particularly useful after students have covered the content in the main textbook. Encouraging students to pose questions about the text will help to uncover the main points.

For example:

- Why did the author choose this topic?
- What does this graphic novel tell me about the people we have studied? Does the story relate ideas about their society, culture, religion, government, military, or economy, to other aspects of their life?
- Is the tone of the story humorous or serious?
- Do I like the people being presented?
- Does the author portray the characters in a positive or negative way?
- What conclusions do these ideas suggest?

Have students write a few sentences answering these questions. Then have them summarize what they believe is the main point of the graphic novel.

6 Visualizing. Have students skim the chapter or a particular section of the chapter. Students should then pick one person, one event, or one concept from their reading and create their own graphic representation of it. Students could use a comic book style to illustrate their topic. Their work could be funny, sad, serious, satirical, or any other tone that they wish. They can use text and dialogue or let the pictures alone tell the story. Another option would be to use other forms of media for their depiction of the topic. Students could take pictures, make a computer slide-show presentation, make a video, or create a song to represent their topic.

These are just some of the many uses of graphic novels. As you introduce them into your class, you may discover more ways to use this popular form to engage your students in a new method of learning while exercising the multiple literacies your students already possess. We welcome you to the world of learning through graphic novels!

Fisher, D., & Frey, N. (2004). *Improving adolescent literacy: Strategies at work.* Upper Saddle River, NJ: Merrill Education.

Frey, N., Fisher, D., & Hernandez, T. (2003). "What's the gist?" Summary writing for struggling adolescent writers. *Voices from the Middle, 11*(2), 43–49.

Geography in Graphic Novels

The following pages contain additional information about each graphic novel. You will find background information, a brief summary, and two activities to help you guide your students' understanding of each graphic novel. The first activity is designed to help the student utilize the story presented to complete the assigned task. The second activity is more broadly focused, allowing students to make connections between the graphic novel and the larger historical context of the period.

Unit 1, Story 1

View from Above

Summary

In 1993, after years of competing for outer space dominance, the United States and Russia decided to work together and build the International Space Station (ISS). In 2000, the first team of inhabitants, Russian cosmonauts and American astronauts, moved in for an extended stay. Since then, crew members from over a dozen different countries, have conducted various types of research intended to improve the quality of life here on Earth. Station crews have studied how cancers grow and have worked with antibiotics to find faster ways to produce them on Earth. They've grown plants to develop drought-resistant crops and crystals to improve gasoline production. Other experiments were designed to take advantage of the low gravity on the space station to study certain physical processes. By eliminating the force of gravity, researchers can better understand some of the smaller forces that occur in processes like semiconductor production.

The astronauts and cosmonauts who live and work on the ISS must undergo a minimum of 18 months training before they qualify for an assignment. They must also speak at least two languages, preferably Russian and English so that they can communicate easily with both the Russian Mission Control Center and NASA's Mission Control Center. While on board the space station, crew members keep a rigorous schedule of work and leisure that includes 2½ hours of exercise a day, which they need to maintain muscle tone and overall fitness in their weightless environment.

This graphic novel takes place aboard the International Space Station as it circles 220 miles above Earth's surface. As the space station orbits above Earth, crew members gaze out the window at the world below and comment on the physical features of their respective homelands and reminisce about the people who live there and the places they miss on Earth.

Activities

Create a physical map showing what you think North America, Europe, South America, Africa, or Asia would look like if you were looking down on it from the International Space Station. Use colored pencils, crayons, or markers to illustrate the various land forms you would be able to see. Include rivers, lakes, seas, mountain ranges, and deserts.

Ask students to think about what it must be like to live on the International Space Station for months at a time. Then, have them imagine that they have the opportunity to interview someone who had done so. Have students prepare a list of questions they would like to ask that person. Now, have students do the research needed to answer their own questions, using the Internet, books, or magazines. Finally, have them write the answers to their questions as though they were being answered by the person being interviewed.

Unit 2, Story 1

Renaissance Man

Summary

Through the concept of a newspaper interview, this graphic novel explores some of the accomplishments of Renaissance genius Leonardo da Vinci. Da Vinci was born in 1452 in northern Italy, the area where the Renaissance began, into an average family. As a child he loved to roam the Tuscan countryside. He was fascinated by lizards, crickets, snakes, butterflies, plants, and flowers. He studied them closely and drew many pictures of them. When his father noticed his talent for drawing, he arranged for Leonardo to become an apprentice to the established artist Verrocchio.

During the Renaissance, merchants who became wealthy from trade hired artists to live on their estates and create art for their large and lavish homes. This practice was called the patronage system and freed artists from the worry of having to make a living. They could devote all their time to making art. The patronage system made it possible for the arts to achieve phenomenal growth during the Renaissance. In Leonardo's case, he was often asked to do a lot more than create art. Although well known for painting the *Mona Lisa* and *The Last Supper*, da Vinci considered himself an engineer more than an artist and spent most of his time designing projects to benefit cities.

Throughout his life, Leonardo da Vinci kept extensive notebooks in which he wrote down his thoughts, ideas for projects, sketches of things or interesting faces of people, along with letters, music, and more. From his notebooks, we know that Leonardo had a natural curiosity about everything, had an unquenchable thirst for knowledge, and was always thinking of ways to do things better. Many modern behavioral specialists believe that these are the very qualities and behaviors that create genius.

Activities

Suggest that students create their own notebook from construction paper or plain white typing paper. Then for the next week have them jot down their ideas for new inventions or improvements they would make on tools or common household objects, ideas for songs, sketches for drawings or paintings they would like to do, a plan for a flower or vegetable garden, or any other project that comes to mind. Then at the end of the week, have them go through their notebook and review what they have written. They should be amazed by the sheer number of creative ideas that come to them.

During the Middle Ages, most art was religious in nature and the techniques used by artists did not result in colorful, lifelike images. Renaissance artists changed all that and leaned more toward the thinking of Roman and Greek artists in approaching their subject matter. Have students research Renaissance art and write a paragraph comparing and contrasting Renaissance art with the artistic style of the Middle Ages.

Unit 2, Story 2

Escape from East Berlin

Summary

As economic conditions in East Germany worsened in the years following World War II, more and more East German citizens decided to leave their country for life in the more prosperous and non-Communist West Germany. In the first seven months of 1961, more than 207,000 people emigrated from East Germany, including hundreds of professionals such as doctors, dentists, and engineers. The East German government was frantic to stop the exodus. So, on Sunday night, August 13, 1961, while people were asleep, East German and Soviet soldiers built the Berlin Wall. Practically overnight, the city of Berlin was divided, splitting up families and friends and preventing East Germans who worked or went to school in West Berlin from continuing to do so. Tearful citizens gathered in the streets fearful they would never see their friends or family again.

In those first days after the wall was built, East Berlin citizens continued to escape by jumping from the upper-story windows of buildings that bordered the Berlin Wall into fire nets held by West Berlin police officers. The windows were soon sealed. Others escaped by boldly climbing over the barbed wire and running to freedom.

To stop the escapes, the East German government built a “death strip” more than half a mile wide ringed by alarms, barbed wire, mine fields, and guard towers. Armed guards and attack dogs patrolled the area. During the years the wall was in place nearly a thousand men, women, and children lost their lives trying to escape. Tens of thousands were caught and imprisoned. Still, people continued to devise ingenious methods of escape using hot air balloons, gliders, and tunnels.

This graphic novel tells the story of one family's daring attempt to leave East Berlin. If caught, they know they will spend the rest of their lives in prison. Their plan is to retrofit, or reduce the size of their car's engine parts, so that there would be room for their grandson to hide inside the engine compartment of the car. They must also outwit vigilant border guards and the guards' well-trained attack dogs.

Activities

Ask students to interview a member of their family or another adult who would remember the Cold War era and get an impression of the international tensions that existed during that time. Students might ask questions about memorable escape attempts from behind the Berlin Wall, what it was like to live during the Cold War era, or what made people willing to risk their lives to escape from Communist countries. Have each student compile the responses in a short report and present it to the class.

Have students study a map of Europe as it existed in 1960. Then have students fold a sheet of paper in half. On one side they should list all of the European countries that were aligned with the Soviet Union and on the other half list all of the countries that were aligned with NATO, or the North Atlantic Treaty Organization, and the United States. Discuss why the countries were aligned that way. What has changed since the 1960s?

Unit 3, Story 1

A Victory for General Winter

Summary

West of the Ural Mountains, where most of Russia's people live, the climate offers warm and rainy summers, but the winters are snowy and brutally cold. Adolph Hitler and his German army set out to invade and conquer the Soviet Union in June 1941. They were not the first to try. Napoleon and his army tried to conquer Russia in 1812 and failed miserably. The cold Russian winter forced the French to retreat. Few of Napoleon's troops made it home to tell their story.

Using a personal diary as a means of expression, this graphic novel follows the footsteps of a young German officer, a geography teacher in his civilian life, as he and his troops march through the European countryside on the way to Moscow. The German soldiers expected to surprise the Soviet army, win a quick victory, and be home by Christmas. The summer months passed easily. As the Germans approached Russia's major cities, the fall weather turned cold and rainy. Supplies got bogged down in the muddy terrain. Soldiers became sick and hungry. To make matters even worse for the Germans, the Soviet people pursued a “scorched earth policy” by killing their livestock and burning everything that could be of value to the Germans.

Winter came early with heavy snow and the coldest temperatures in years, dropping to –40°F. Many German troops were still in summer uniforms and died from cold and frostbite. Others starved to death. The severe cold froze trucks, tanks, and airplanes, rendering them useless. Meanwhile, well-fed and warmly dressed Soviet troops took the offensive. Accustomed to the cold, and with reinforcements from Siberia, they stopped the Germans outside of Moscow and forced them to retreat.

Once again, the cold Russian winter helped repel foreign attackers and sent them home haggard and defeated.

Activities

In the graphic novel, ask students to find instances in which the German soldier mentions supplies. Then have each student write a paragraph addressing the following questions: Why do you think the Soviet people killed their livestock and burned their houses and crops? How does Russia’s geography lend itself to such a strategy? What would you have done if you were in their situation?

Using the maps in the front of their textbooks, have students make a list of the countries that the German army would have passed through on the way to Moscow. Have them name the capital of each country and briefly explain the terrain the army would have passed through and any major rivers they would have crossed. Approximately how many miles did the German army travel if they started in Berlin?

Unit 3, Story 2

Masha and the Bear, A Russian Folktale

Summary

Russia has a vibrant cultural history, due in part to the contributions of the many ethnic groups that live within the country. Storytelling is one of the oldest and richest of Russia’s cultural traditions. Originally passed down orally from generation to generation, when people spent long, dark winters in front of their fireplaces, the stories eventually were collected, printed, and illustrated. The stories often include animals with mystical powers, beasts, and people facing moral decisions.

This graphic novel takes an old Russian folktale and introduces it to the reader with a modern twist. A teenage girl approaches her parents, who are in the company of her grandfather, asking to go to the mall with her friends. When her parents question her about who is driving and when she will be home, she becomes defensive and irritated. Her grandfather steps in as mediator, and as a way to illustrate that the parents aren't being unreasonable, he tells the girl an old Russian folktale. The folktale is about Masha, a Russian girl who doesn't follow her family's instructions to be safe and ends up being captured by a bear who won't let her go home.

Activities

Ask the students to change the ending of the story. They can rewrite it by creating their own escape plot for Masha, by having her rescued by another person or animal, or perhaps by having Masha continue to live in the woods with the bear until she is very old.

Using various resources including magazines and newspapers, have students collect information on art, music, dance, and literature in Russia and then use it to create a cultural scrapbook. Students can either focus on one aspect of the arts in Russia, such as ballet, or they can present a survey of material from all of the artistic disciplines. Material can be mounted on construction paper or lightweight poster board.

Unit 4, Story 1

Cat Mummies

Summary

In this graphic novel, two young Egyptian girls spend an afternoon with their grandfather, a retired anthropology professor, at a rather unusual museum. The museum is unique because it is dedicated to the memory and mystique of cats in ancient Egyptian history. In the special exhibits, the girls are treated to a variety of images and artifacts relating to the lofty position cats held in Egyptian history. They learn why cats were held in such high esteem and what happened to them after they died. Of particular interest to the girls is a painting just inside the front door that shows the museum's founder with her beloved pet cat. This beautiful cat wears a distinctive collar and has eyes that seem to follow the girls' movements.

The Egyptians believed that, like people, cats had a spiritual twin called the *ka* that was created when they were born. They believed that when they died the *ka* would come back to the body and give it new life. But in order to do so, the *ka* had to be able to recognize the body to which it belonged. To ensure that it did, the Egyptians preserved, or mummified, the body.

Activities

Explain to students that when the cats of wealthy owners died, their owners had the cats' bodies wrapped in strips of dyed linen, which were woven into beautiful patterns. Decorated papier-mâché masks covered their heads. Many were placed in wood or bronze coffins, some of which were shaped like cats. Burial took place in cemeteries just for cats. On sheets of paper or poster board, have students design and draw pictures of cat mummy cases and decorate them using colored pencils, crayons, or markers.

Ancient Egyptians made contributions in many areas besides mummification. Using their textbooks as a reference, have students choose another contribution the Egyptians made to civilization. Have students expand on the information they find in their textbook by researching books, magazines, and the Internet. Have them compile a short report about their chosen subject.

Unit 4, Story 2

Nebuchadnezzar's Babylon

Summary

King Nebuchadnezzar conquered most of the lands and people of the Fertile Crescent before settling down to govern from the ancient city of Babylon. The city was situated on the Euphrates River, just south of where Baghdad, Iraq, is located today. The name Babylon means "gate of the gods." During his reign, Nebuchadnezzar rebuilt and expanded the centuries old city into a wealthy center of trade and culture. Historians wrote that the city's gates, elaborate palaces, and temples glittered with glazed brickwork in intense colors of red, yellow, blue, and cream. The city also featured the world's first masonry bridges, 400 feet long, which spanned the Euphrates River and connected the two halves of the city. Most of the estimated population of 150,000 lived in small brick houses that faced away from the street, toward courtyards, both for privacy and for shade from temperatures that could reach 130°F.

In this graphic novel, King Nebuchadnezzar proudly guides the renowned Greek historian, Herodotus, on a tour of his cherished city of Babylon. Ancient historians noted that Babylon was in a state of decay when Nebuchadnezzar came to power and that his accomplishments in rebuilding the city were more important to him than all of his military victories. The king also built a museum to house all of the artifacts from earlier civilizations that were found during renovation and excavation. Unfortunately, after his death, the city deteriorated quickly and disappeared, until ruins were discovered in the early 1800s.

Activities

As they read the story, have students imagine that they are on vacation accompanying Herodotus and King Nebuchadnezzar on a tour of Babylon. Then ask the students to write a letter home to a friend or family member describing what they saw on the tour and giving their overall impression of the city of Babylon.

Civilization existed and cities thrived in the Fertile Crescent centuries before King Nebuchadnezzar rebuilt Babylon. During that time, the people who lived in the region made many discoveries and invented many items that are still in use today. Have students research and list some of those discoveries and inventions, choosing one to write about in more detail. If time permits, have students share their reports with the class.

Unit 5, Story 1

The Vanishing Gorillas of Central Africa

Summary

Have you ever imagined what the world looks like through the eyes of an animal? Have you ever wondered how animals perceive human behavior? This graphic novel looks at events through the eyes of mountain gorillas, one of the most endangered species of wildlife in the world. Only about 700 mountain gorillas exist. Mountain gorillas live in the cloud-shrouded mountains of the Virunga range in the Democratic Republic of Congo, Rwanda, and Uganda. Their lives are constantly threatened by habitat loss, poaching, and civil war.

Mountain gorillas have a well-developed social structure. Each group contains a dominant male, a number of adult females and their offspring. There is an overlap in group territories and the dominant male generally defends his group rather than a specific territory. High infant mortality, a long gestation period (251—295 days), a tendency for single births, and a prolonged period of maternal care are all factors that contribute to low population growth. Females generally give birth to only two or three gorillas during their entire lifetime which averages about 50 years.

Although strong and powerful, mountain gorillas are generally shy and peaceful. When threatened, the male will beat his chest, pound his fists on the ground in an aggressive manner, and even charge, but he will seldom strike out. Mountain gorillas eat a mainly vegetarian diet, and they will spend half their day feeding on stems, bamboo shoots, berries, fruits, and insects.

The mountain gorilla is threatened by humankind. Poachers kill entire groups to capture infants for zoos and the illegal pet trade and to acquire heads and hands for trophy hunters. The strongest threat to gorillas' existence comes from deforestation and habitat loss as the rapidly increasing human population in the area clears forest land for firewood and farming. Ironically, it may be ecotourism that saves the mountain gorilla. As an ever increasing number of European and American tourists visit the area, awareness of the gorillas' situation spreads and more people commit both time and money to save the gorillas from extinction.

Activities

To learn more about mountain gorillas, their diminishing habitat, and the International Gorilla Conservation Program, have students go to the African Wildlife Foundation at www.awf.org or the World Wildlife Fund at www.worldwildlife.org/gorillas/subspecies/mountain.cfm. Then, have each student write a persuasive paragraph explaining why he or she thinks that people around the world should care about what happens to gorillas in Africa. Students should list at least two arguments that support their opinion.

Explain to students that the Democratic Republic of the Congo is one of the three countries where mountain gorillas live. It is a large country with a wealth of undeveloped natural resources. Using their textbooks as a reference, ask students to list the resources that are available for development. Have students write a paragraph that summarizes why they think the country is not taking advantage of its mineral wealth.

Unit 5, Story 2

South Africa: A Time for Change

Summary

Although South Africa is the most industrialized and most prosperous country in Africa, until recently, only a small minority of the population was able to benefit from the country's wealth. Most of the population, about 78 percent, is made up of people from black ethnic groups including the Zulu and Sotho who have been kept out of the mainstream since apartheid became law in 1948. Under apartheid, it was illegal for different races and ethnic groups to mix. Black South Africans were forced out of the cities and suburbs and sent to live in rural areas called townships. Most townships lacked basic services such as electricity and running water. People who came from non-European ancestry were not allowed to vote and could only hold menial jobs.

For nearly 40 years people around the world protested against the practice of apartheid. Many nations refused to trade with South Africa and discouraged their citizens from visiting the country or participating in sporting events against South African teams. Many black people in South Africa lost their lives in the struggle for freedom and many others were jailed for speaking out. Apartheid finally came to an end in 1991. In 1994, for the first time in history, white and black South Africans were able to vote in a national election. Nelson Mandela, a man who spent 27 years of his life in jail for speaking out against apartheid, was elected the country's first black president.

This graphic novel follows two South African citizens, one black, one white, through a seemingly typical weekday that culminates in an extraordinary event in which both men, one for the very first time, cast their vote for a new president.

Activities

Ask the students to compare and contrast the images in the novel that show how the two South Africans live their lives. Look at jobs, homes, transportation, etc. Then have the students list their impressions on a sheet of paper that has been folded in half, with one side devoted to the lifestyle of the white South African and the other side devoted to the lifestyle of the black South African. When the students are finished, ask them to write a short paragraph explaining why the two lifestyles are so different and given a choice, which one they would prefer to live.

The Republic of South Africa, larger in area than the states of Texas and California combined, lies at the very tip of Africa. The country is bordered by two oceans and has a diverse physical structure. Have students create a map of South Africa that includes its major geographical features such as deserts, mountain ranges, rivers, lakes, forests, and the location of its major cities. Be sure to include the oceans and indicate bordering nations.

Unit 6, Story 1

Sherpas to the Rescue

Summary

The kingdom of Nepal is the doorway to eight of the tallest mountains in the world. Farming is the primary source of income for 80 percent of the people in the country, but income from tourism is growing. One ethnic group in Nepal, the Sherpas, are a poor people who earn their living by subsistence farming and by guiding hikers and climbers through the Himalaya. Despite their hard work and the heroic acts they often perform to help those who climb Mount Everest for fame and adventure, Sherpas seldom receive any recognition and are not greatly rewarded financially. Tenzing Norgay, who reached the summit with Sir Edward Hillary in 1953, is one of the few Sherpas to receive any sort of acclaim.

This graphic novel tells the story of two American tourists who seek adventure by attempting to climb Mount Everest, which at 29,035 feet is the world's tallest mountain. The two adventurers approach the climb up the mountain with much bravado, touting their physical and emotional prowess. Meanwhile, their Sherpa guides, amused by the tourists' attitudes, make the climb carrying all of the group's climbing equipment and camping supplies. They also set up and take down the camp each day as well as cook the meals and do the dishes. After days of climbing, our expedition finally reaches the summit of the mountain and our tourists launch into a frenzied celebration over this extraordinary feat they performed "all on their own."

Activities

When Tenzing Norgay and Sir Edward Hillary became the first men to reach the summit of Mount Everest in 1953, they stayed at the top of the mountain only a very short time because of the lack of oxygen at that height. Today's climbers carry bottled oxygen which enables them to stay at the summit for longer periods of time. Have students research and make a list of other modern conveniences, some developed by NASA for use in space, that make climbing the mountain easier today than it was for Norgay and Hillary.

Suggest that students create an advertisement that will persuade people to come to Nepal for their next vacation. It can be designed for newspaper, magazine, television, radio, or even the Internet. Have students use their text-books, the Internet, an encyclopedia, or other books and magazines to conduct research. Students can either clip pictures from magazines or draw their own.

Unit 7, Story 1

Protest on the Square

Summary

China's civilization dates back more than 4,000 years. For much of its history, China was ruled by a series of emperors and empresses. In 1911, the last emperor was overthrown in an uprising and China became a republic. After

World War II, the Nationalist Party and the Communist Party fought for control of the country. The Communist Party won. Since 1949 China has had a Communist government and has been called the People's Republic of China.

China's Communist government does not tolerate criticism. Certain basic human rights like freedom of speech and freedom of the press are not available to China's citizens. Anyone who speaks out for more freedom or criticizes China's current government can be jailed or even killed.

In 1989, about 100,000 students and workers gathered in Beijing's Tiananmen Square for a peaceful political rally in support of free speech and a free press. They also wanted China's aging leaders, who hold their jobs for life, to step down. The dissenters timed their protest to coincide with a visit to their country by Mikhail Gorbachev, the leader of the Soviet Union. Realizing that all of the international news networks would be covering Gorbachev's visit, the protestors hoped the publicity would focus the world's attention on their cause. This graphic novel explains what happened to the protestors and the movement for more freedom in China.

The peaceful political rally turned violent. China's Communist leaders sent armed military troops and tanks into the square. Chinese soldiers fired at random into the crowd, killing hundreds and wounding thousands. The remaining protestors were frightened into silence, and foreign news crews were sent home.

Activities

Ask the students to think about what might have happened if Mikhail Gorbachev and the leaders of other Communist countries had appealed to the Chinese leaders to soften their attitudes on dissent and let the protestors present their views. Have the students rewrite and illustrate the end of the graphic novel accordingly.

The Chinese government has been criticized around the world for its harsh treatment of its citizens. Many people have suggested that other nations stop trading with China, a measure similar to the boycott of South Africa during apartheid, until the Chinese government starts giving its citizens basic human rights. Ask each student to write a paragraph explaining what is meant by the term human rights. In a second paragraph, have each student give his or her opinion on whether a trade boycott would be a fair and effective way to influence China's stand on human rights. If students approve of such a boycott, have them explain how they might show their support for the measure.

Unit 7, Story 2

Divine Wind

Summary

Kublai Khan and the Mongols enjoyed great military success. After completing his conquest of Southern China, Khan focused his attention on Japan. The Mongols were fierce competitors, but they failed on two occasions to conquer the Japanese, first in 1274 and then again in 1281. Japan's geography and weather played a significant role in the Mongol defeat.

In the first attempt, Khan brought 900 ships and was experiencing some success, but was turned back by poor weather conditions. Khan was better prepared for the second attempt, bringing over 3,500 ships. This attempt failed when a typhoon struck. Khan had ordered his ships chained together to protect them from being boarded and attacked by Japanese samurai. When the typhoon hit, the ships were trapped in the shallow waters around Japan and could not make it out to the safety of deeper water. The wind and waves smashed them against each other and the rocky coast. Many were sunk.

The Japanese credited this victory to their gods. They called the typhoon *kamikaze* or "divine wind." The term was later used to describe Japanese suicide bombers during World War II.

Activities

Ask students to surmise what might have happened if a typhoon had not struck when it did. Would the Japanese have been able to protect their island from the invaders without the Divine Wind? Have students rewrite and illustrate the end of this graphic novel to reflect their opinions.

The islands that comprise Japan are actually the tops of rugged volcanic mountains that rise from the ocean floor. Have students look at a map of Japan and point out to them the major geographical features of the country. Ask students to consider how Japan's geography would affect any attempted invasion and ask them to write a paragraph explaining their views.

Unit 8, Story 1

Island Time

Summary

This story takes place on one of the 25,000 islands that comprise Oceania. Some of the islands originated from volcanic eruptions; others were built up by coral. The region offers a hospitable climate with temperatures that range between 70°F (21°C) and 80°F (27°C). The surrounding seas are rich in marine life, and volcanic soils produce bananas, coconuts, sweet potatoes, and cassava.

In this novel, a Hawaiian middle school teacher offers his students the opportunity to spend the summer on a distant island where they will use the knowledge and survival skills he taught them during the school year to live off the land without using any modern amenities. The students board a double-masted schooner which takes them to their island destination. There, they are left on their own to build shelters, acquire food, and distribute the chores among themselves. Just when they think they are totally on their own, a local family pays the students' camp a visit and offers to show them how native islanders use the island's resources to their advantage. The students have a productive and fun-filled summer, and on their last day a masked visitor arrives with a little secret to share.

Activities

Have students look through the graphic novel and note any tools or other useful items that the visitors are able to create from natural products on the island, for example, their shelters, fire, fishing spears, fishing nets, and an oven. What items does it appear that the visitors brought to the island? Then, ask students to think about what they would pack and take with them if they were going to be on an island for several months. Have them make a list of the items they would want to have with them. Would they want to take a hammock, a supply of chocolate bars, or their favorite book? Discuss some of the students' choices and the reasons they would take those items.

The islands of the Pacific are either high volcanic islands or low islands, most of which are atolls made of coral. Have students research, build, and label models of high and low islands using a variety of materials including modeling clay, papier-mâché, cardboard, construction paper and so on. Suggest that they add enough detail to indicate the specific types of vegetation and resources that would be found on each type of island.

Unit 8, Story 2

Shackleton's Antarctic Adventure

Summary

This graphic novel recounts one of the most awesome man-against-nature sagas ever recorded. In 1914, Sir Ernest Shackleton, a British explorer, and his 27-man crew set out for Antarctica aboard a ship named the *Endurance*. The *Endurance* was only one day away from reaching the continent when it became trapped in sea ice. For the next 10 months, the ship was frozen in the ice, despite the crew's efforts to free her. Eventually the increasing pressure of the ice crushed the ship's hull and the ship sank, leaving her crew stranded on an ice floe in one of the most inhospitable places on Earth.

The following months would be a testament to Shackleton's extraordinary leadership skills, as well as to the resourcefulness and camaraderie of the crew members. Before the ship sank, the crew rescued the three life boats, tents, sleeping bags, and other supplies. They left most of their personal possessions behind. For five months the crew camped on the ice, braving the fierce cold and dwindling food supplies. When the ice started to break up, Shackleton made a decision to load everything into the life boats and sail to Elephant Island and firmer ground.

With hope of rescue dimming, Shackleton again made a critical life or death decision. It seemed the crew's only chance of survival was to get help at a whaling station on South Georgia Island, a treacherous 800-mile journey across the stormiest seas in the world. So, while most of the crew remained on Elephant Island, Shackleton and a few of his men set sail in a 23-foot open boat on an eight day journey across the open sea. They suffered through gales, thirst, and extreme hunger. Even when they reached the coast of South Georgia Island, their journey was not over. The men still had to hike across an uncharted mountain range to reach the whaling station on the other side of the island. Even with additional help and supplies, it would take three separate attempts, each thwarted by ice-choked seas, to reach the remaining crew on Elephant Island. Shackleton, however, never gave up. On August 30, 1916, nearly two years after first arriving in Antarctica, Shackleton reached his remaining crew. Miraculously, everyone survived.

Activities

Have students review the following web sites for more detailed information on Sir Ernest Shackleton and the crew of the *Endurance.*

American Museum of Natural History at www.amnh.org/exhibitions/shackleton/ (Click on "expedition" for a detailed chronology of the journey.)

NOVA Online at www.pbs.org/wgbh/nova/shackleton (includes links to a crewmember's diary)

Once students have reviewed the information on the sites, have them imagine that they are newspaper reporters who have the opportunity to interview Shackleton or one of his crew. Ask students to make a list of questions they would want to ask. Then have students write answers to their questions based on the information they found on the sites. Share both the questions and answers with the class.

According to the terms of the Antarctic Treaty, the nations of the world have agreed to use Antarctica only for peaceful, scientific research. One of the most important long-term research studies being conducted in Antarctica is focused on the deteriorating ozone layer. Have each student write a paragraph giving at least two reasons why the results of this research might be important in the future.

Unit 1

The World

SAM COTTON. GREAT FALLS, MONTANA
I'VE BEEN ON THE SPACE STATION FOR A FEW MONTHS NOW, BUT I USED TO LIVE IN BIG SKY COUNTRY. YOU KNOW, ON EARTH.
THERE'S NOTHING LIKE IT DOWN THERE. IT'S SO BEAUTIFUL, WITH ITS ROLLING HILLS AND ITS WIDE RIVER VALLEYS.
I'VE GOT A PHOTO HERE. THAT'S BIG SNOWY. THE MOUNTAIN RANGE IS CALLED BIG SNOWY. THE HORSE...WELL, HIS NAME IS FRED.

ADRIKA KALYANI.
MUMBAI, INDIA
WHEN I LOOK DOWN ON MY HOME COUNTRY IT'S HARD TO IMAGINE THAT THERE ARE MILLIONS OF PEOPLE LIVING THERE.

RIGHT NOW, DOWN THERE, IN INDIA, THOSE MILLIONS OF PEOPLE ARE EXPERIENCING THE MONSOON SEASON AFTER A VERY DRY WINTER.
IT'S ONE OF THE MOST DRAMATIC CLIMATE PHENOMENA ON THE PLANET.

THAT RAINFALL IS ESSENTIAL TO CROP GROWTH.
ALL THOSE PEOPLE NEED TO BE FED, YOU KNOW.

VERONIKA KOTOV.
IRKUTSK, RUSSIA
SOMETIMES I FEEL SO CLAUSTROPHOBIC ON THIS STATION.

I'M FROM THE LARGEST COUNTRY IN THE WORLD-- *MOTHER RUSSIA!*

"IT SPREADS ACROSS TWO CONTINENTS-- EUROPE AND ASIA.
"IT'S SO BIG, THAT IF YOU TRAVELED FROM EASTERN RUSSIA TO WESTERN RUSSIA, YOU WOULD PASS THROUGH ELEVEN TIME ZONES.

BLAH! I FEEL LIKE A SARDINE IN THIS FLOATING TIN CAN.

EMIEL HUYGELEN.
ANTWERP, BELGIUM
I MISS FISH. I MISS FISHING TOO.
MOSTLY, I MISS MY HOME.

"DID YOU KNOW, THAT DESPITE ITS SHORT COASTLINE ON THE NORTH SEA, BELGIUM IS A PRINCIPLE TRANSPORT AND TRADE CORRIDOR IN EUROPE?

IT HAS INCREDIBLE SEAPORTS, YOU SEE.
THE DOCKS ARE ALWAYS BUSTLING.

YOU KNOW WHAT I REALLY MISS? I MISS THE EXCITEMENT.

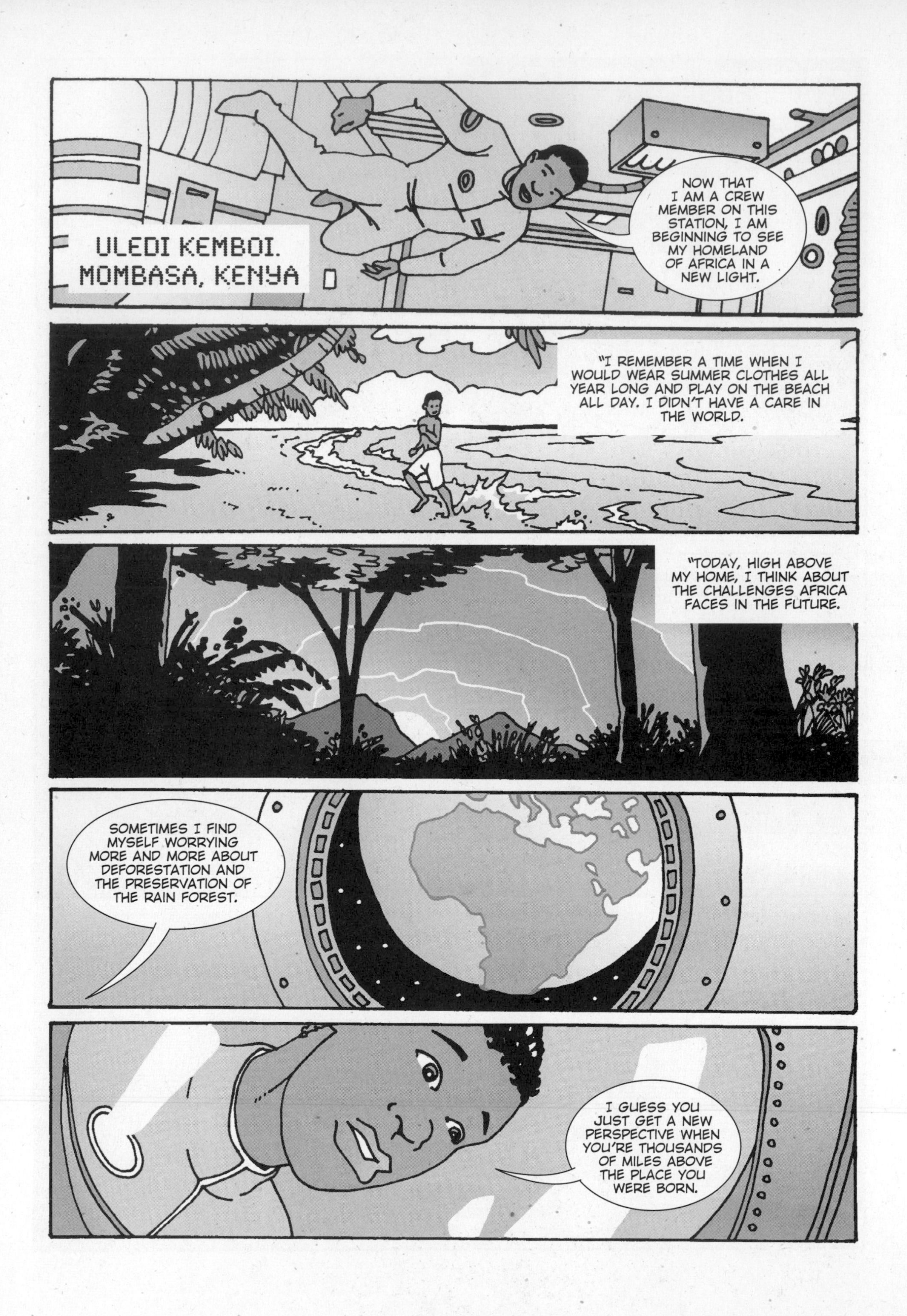
ULEDI KEMBOI. MOMBASA, KENYA
NOW THAT I AM A CREW MEMBER ON THIS STATION, I AM BEGINNING TO SEE MY HOMELAND OF AFRICA IN A NEW LIGHT.
"I REMEMBER A TIME WHEN I WOULD WEAR SUMMER CLOTHES ALL YEAR LONG AND PLAY ON THE BEACH ALL DAY. I DIDN'T HAVE A CARE IN THE WORLD.
"TODAY, HIGH ABOVE MY HOME, I THINK ABOUT THE CHALLENGES AFRICA FACES IN THE FUTURE.
SOMETIMES I FIND MYSELF WORRYING MORE AND MORE ABOUT DEFORESTATION AND THE PRESERVATION OF THE RAIN FOREST.
I GUESS YOU JUST GET A NEW PERSPECTIVE WHEN YOU'RE THOUSANDS OF MILES ABOVE THE PLACE YOU WERE BORN.

EVERYONE OF US ABOARD THE SPACE STATION CALLS A DIFFERENT PART OF EARTH HOME.
LISTENING TO MY NEW FRIENDS REFLECT ON ALL THE INCREDIBLE PLACES THEY'VE SEEN AND HEARING ABOUT WHERE THEY'VE LIVED, I FEEL EVEN CLOSER TO THAT BEAUTIFUL SPINNING BALL OUT THERE.

YEAH, THE EARTH MAY BE AN AMAZING SIGHT FROM UP HERE, BUT I CAN'T WAIT TO GET BACK DOWN THERE AGAIN AND EXPERIENCE IT UP CLOSE.
DANGER
ОПАСНОСТЬ
GEFAHRENGEBIET

THE END.

Unit 2

Europe

FLORENCE, ITALY. CIRCA 1510.
MR. DA VINCI, ON BEHALF OF THE FLORENCE TIMES, I MUST THANK YOU AGAIN FOR AGREEING TO BE INTERVIEWED.
OH, WELL, IT IS AN HONOR. I WILL APOLOGIZE IN ADVANCE IF I'M A LITTLE DISTRACTED. YOU SEE, I DO LIKE TO WORK ON A VARIETY OF PROJECTS, ALL AT THE SAME TIME.
CAN YOU GIVE US A BRIEF SUMMARY OF WHAT YOU'RE WORKING ON TODAY?
HMM. WELL, OFF THE TOP OF MY HEAD...
A PORTRAIT IN OILS FOR A PATRON, A WATER PUMPING MACHINE I'M INVENTING, SOME FANCIFUL MANNED FLYING MACHINES, AN ESSAY ABOUT ART, SOME LETTERS I NEED TO FINISH WRITING...
YES, THAT'S ALL I CAN THINK OF FOR TODAY...
WOW! THAT CERTAINLY IS A LOT! ARE THOSE PAPERS BEHIND YOU RELATED TO THESE PROJECTS?

THIS MESS? OH, YOU SHOULDN'T EVEN BE LOOKING AT THESE, YOUNG MAN! MY PATRONS MIGHT DISAPPROVE!
PERSONALLY, I DON'T MIND. AS A MATTER OF FACT, IF YOU'D LIKE, YOU CAN TAKE A CLOSER LOOK AT MY NOTEBOOK.
WHAT? REALLY?
OF COURSE. IT'S ALSO MY WAY OF TURNING YOUR ATTENTION AWAY FROM THE WORK I'M DOING FOR MY PATRONS.
HA HA!
WOW, MR. DA VINCI, THIS NOTEBOOK IS AMAZING!
PLEASE, CALL ME LEONARDO.
IN THAT NOTEBOOK, YOU'LL FIND EVERYTHING FROM LIFE DRAWINGS TO ENGINEERING DIAGRAMS TO VARIOUS WRITINGS ABOUT MY NUMEROUS INTERESTS.

INCREDIBLE...
THAT REMINDS ME, I REALLY DO HAVE TO GET TO WORK ON THIS CATAPULT DESIGN...

I'M SORRY, BUT CAN WE CONTINUE OUR INTERVIEW TOMORROW?
OH, BUT MY DEADLINE—

DEADLINES CAN WAIT, I ASSURE YOU. FOR YOUR COOPERATION, YOU ARE FREE TO BORROW THAT NOTEBOOK UNTIL THEN IF YOU WANT...
I STILL CAN'T BELIEVE IT. I'M HOLDING IN MY HANDS THE WORK OF A TRUE GENIUS.

HALF OF IT I CAN'T DECIPHER; MUCH OF IT IS WRITTEN BACKWARDS.
HMM. THIS IS AN INTERESTING PAGE. STUDIES OF HORSES AND SOME WRITING...

"WE ARE DECEIVED BY PROMISES AND DELUDED BY TIME. AND DEATH DERIDES OUR CARES; LIFE'S ANXIETIES ARE NAUGHT."

WOW.
THAT'S DEEP.

I NEVER REALIZED WHAT A WELL-ROUNDED MAN LEONARDO IS. IN THIS NOTEBOOK ALONE THERE ARE POLITICAL ESSAYS, COUNTLESS ENGINEERING IDEAS, AND WORKS OF MUSIC.

THE AMAZING THING IS THE LEVEL OF INVENTIVENESS AND INSIGHT IN THE DRAWINGS AND WRITING.
NOT JUST ANYONE CAN DO THIS STUFF!

FOR INSTANCE, ON THIS PAGE IS A ROUGH DIAGRAM OF A "FLYING MACHINE."
IMAGINE THAT! SUCH A THING IS COMPLETELY UNHEARD OF IN OUR TIME! BUT THIS DIAGRAM MAKES IT ALL LOOK SO POSSIBLE.

AND HERE, WE HAVE MORE WRITING: "THERE IS NO CERTAINTY WHERE ONE CAN NEITHER APPLY ANY OF THE MATHEMATICAL SCIENCES NOR ANY OF THOSE WHICH ARE CONNECTED WITH THE MATHEMATICAL SCIENCES."
WOW, SO HE'S VERSED IN MATH, TOO!

AND TO THINK THAT MOST OF EUROPE KNOWS LEONARDO JUST AS A PAINTER.

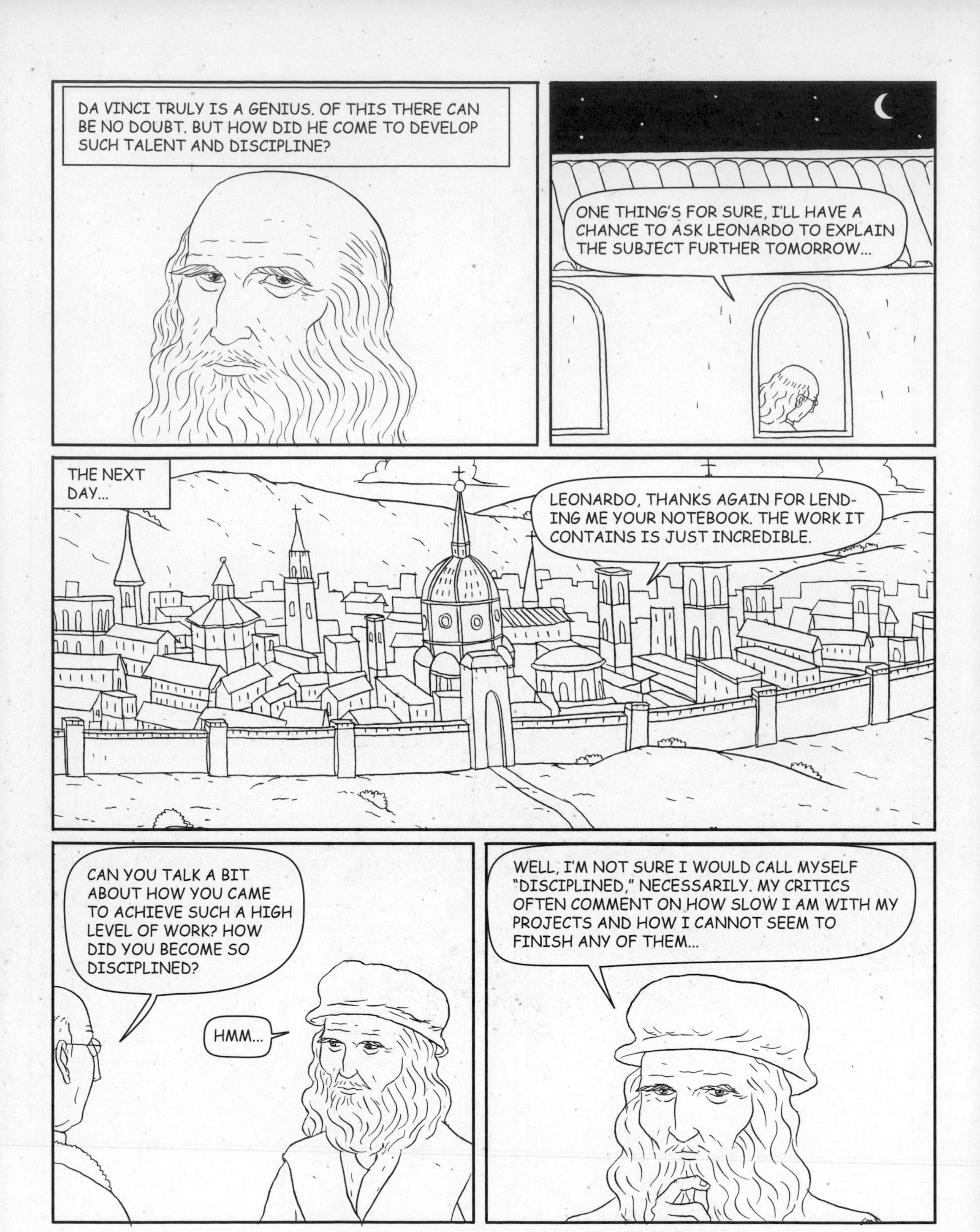
DA VINCI TRULY IS A GENIUS. OF THIS THERE CAN BE NO DOUBT. BUT HOW DID HE COME TO DEVELOP SUCH TALENT AND DISCIPLINE?
ONE THING'S FOR SURE, I'LL HAVE A CHANCE TO ASK LEONARDO TO EXPLAIN THE SUBJECT FURTHER TOMORROW...
THE NEXT DAY...
LEONARDO, THANKS AGAIN FOR LENDING ME YOUR NOTEBOOK. THE WORK IT CONTAINS IS JUST INCREDIBLE.
CAN YOU TALK A BIT ABOUT HOW YOU CAME TO ACHIEVE SUCH A HIGH LEVEL OF WORK? HOW DID YOU BECOME SO DISCIPLINED?
HMM...
WELL, I'M NOT SURE I WOULD CALL MYSELF "DISCIPLINED," NECESSARILY. MY CRITICS OFTEN COMMENT ON HOW SLOW I AM WITH MY PROJECTS AND HOW I CANNOT SEEM TO FINISH ANY OF THEM...

TO THOSE CRITICS I SAY, "BAH!" I AM USUALLY WORKING ON SO MANY DIFFERENT THINGS AT ONCE THAT IT'S DIFFICULT TO FINISH **ONE** THING QUICKLY!

I AM NOT LAZY.

AND AS FAR AS QUESTIONS RELATING TO MY WORK HABITS GO, YOU WILL FIND ALL ANSWERS IN THERE, IN THE NOTEBOOK.

I WORK EVERY DAY, JOTTING DOWN IDEAS, SKETCHES, AND WRITINGS. MY MOST CREATIVE THOUGHTS FLOW FREELY INTO MY NOTEBOOK. AND IT IS FROM MY NOTEBOOK, THAT MY GREATEST ACCOMPLISHMENTS HAVE SPRUNG.

AND FINALLY...

SO THE BEST EVIDENCE OF DA VINCI'S GENIUS LIES IN HIS NOTEBOOKS.

AND WHAT A GRACIOUS MAN! ALL THE RUMORS I HEARD OF LEONARDO BEING MOODY AND GRUMPY WERE GREATLY EXAGERRATED.

YES. DA VINCI TRULY IS A RENAISSANCE MAN.

THE END

Unit 2

Europe

ESCAPE FROM EAST BERLIN

East Berlin, September 1961
Is this all we have for dinner? How can Klaus grow up on so little food?

Be thankful the butcher is your cousin, or we'd have even less!
The shops are empty, like they were during the First World War.

I fear there will soon be another war, and a bigger one at that!
Don't say that, *please!*

I nearly lost you in that war.
And then, in the last one...

Our only son and his beautiful wife...

Now, it's the same as then. The jobs pay nothing. We can't afford what little we find in the shops. And if you dare complain, the Stasi* have their spies everywhere!
Some of my schoolmates are Stasi spies!
**Stasi*: East German Secret Police

Hush, Friedrich! Someone might hear you!

Look at us! We're afraid to speak in our own home! Is this the future we want for our Klaus?
No, but...

Later that week...
We must take Klaus to the West!
But nobody who's young and fit for work is allowed to travel.

And so many have died while trying to escape... by accident or shot by the guards.

But we can learn from their brave attempts and honor them with our success.

I have a plan...

We must act quickly, before they ban *all* travel. We must be ready to leave everything behind. When we are in the West with Klaus, we can never return.

It's the only way my plan can work. We'll never see our friends here again, but our Klaus will grow up a free man. Will you do this with me?

Yes... yes, I will.
For the two of you, my loves.

You and I are retired. We're still allowed to cross back and forth. We'll start doing it more often.
We'll visit Erich, while Klaus stays here. Erich will be able to help us.

Soon...
Grandmother! Grandfather! I've missed you!
We'll be coming to visit every weekend, Erich...

...until the last time, when we'll remain for good.

Erich, are the stories true that people got through the Brandenburg Gate strapped under the frames of cars?
Some stories, yes, but the guard dogs found most of the people.

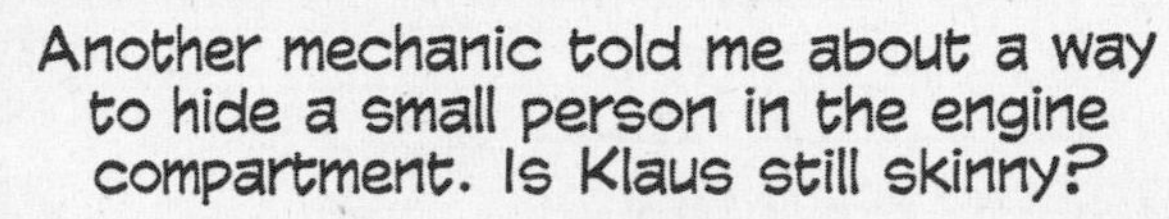
Another mechanic told me about a way to hide a small person in the engine compartment. Is Klaus still skinny?

Ach, yes, too skinny!
Well, Grandmother, don't feed him too much until we have him here in the West.

I can fix the car, but what about the guards and the dogs?
Grandmother and I will disarm the guards.

You and Grand-mother?
That's right! Grandfather has a plan and I think it will work!

Don't worry about the guards. Can you fix the car before Christmas?
I can do a little each week. It will be ready by then.

Eight weeks later...
You two young men are at the gate again? Such fine German boys!
Week after week, they work so hard!

I wish I could give you both cakes for guarding our border so well!
Otto likes cakes!
Ha ha!

Drive carefully, sir!
You're the one who likes cakes!
Whew!

Erich, how soon?
It's nearly ready. I'll finish it in early December.

Perfect. On Christmas Eve, Klaus will ride on the small "shelf" in the engine compartment.
The guards will be distracted by the cold weather and the holiday.

December 24, 1961
You'll be warm in here, Klaus. For goodness' sake, please be quiet.
I will, Grandfather.

Ach, no! My favorite young men working on a snowy Christmas Eve?
Good evening, Mr. and Mrs. Werner.
Grrrr!

Fritz! What are you growling at? You know the Werners. Hush now!
GRRRRRR! RRRRRRRR!

Your dog must smell the Christmas cakes!
Cakes?
Yes, cakes! Happy Christmas, young men.

We really shouldn't take these, but...
It's Christmas Eve, so we will! Be careful crossing the No Man's Zone. The wind is bitter tonight.
Bless you. Stay warm tonight!

It's only a few meters more now, Klaus...

We made it!

Unit 3

Russia

JUNE 4TH, 1941. IN MY SMART NEW UNIFORM, I LEFT BEHIND MY HOME IN WUPPERTAL, MY PARENTS AND GERTA, AND THE LITTLE SCHOOLHOUSE AND HEADED OUT INTO THE WORLD. WE MOVED FIRST ACROSS GERMANY AND THEN INTO POLAND. TRAVELING SOMETIMES IN TRUCKS OR TRAINS, BUT OFTEN BY FOOT.
I AM SURROUNDED BY BOYS BARELY OLDER THAN THOSE I USED TO TEACH. AND, HERE, ON THESE BEAUTIFUL SUMMER DAYS, I CAN IMAGINE THAT WE ARE ON AN EXTENDED FIELD TRIP, A NATURE HIKE.
SOMETIMES I CATCH MYSELF LECTURING THESE YOUNG MEN ON GEOGRAPHY. I'M A LIEUTENANT, SO THEY LISTEN, AND THE POLITE ONES EVEN PRETEND TO BE INTERESTED.
WE'RE MOVING ALONG THE GREAT EUROPEAN PLAIN, CHARACTERIZED BY THESE ROLLING GRASSLANDS AND DECIDUOUS-CONIFEROUS FORESTS...
"CARNIVOROUS FORESTS?"
...MEANING TREES THAT LOSE THEIR LEAVES AND EVERGREENS.

JUNE 22ND. WE HAVE INVADED THE SOVIET UNION. WAS IT ONLY TWO YEARS AGO THAT HITLER AND STALIN SIGNED A NON AGGRESSION PACT? FROM FRIENDS TO ENEMIES, SO QUICKLY.
BUT WHO AM I TO QUESTION OUR GENERALS?
JULY 9TH. THE INVASION HAS BEEN A GREAT SUCCESS SO FAR. MILLIONS OF GERMAN TROOPS POURED OVER THE BORDER, CATCHING THE SOVIETS OFF GUARD. WE MOVED QUICKLY, IRRESISTIBLY, THROUGH THE FLATLANDS OF LITHUANIA...
Gulf of Riga
LATVIA
Riga
BALTIC SEA
Dvinsk
Polotsk
LITHUANIA
Tilsit
Konigsberg
Danzig
Vilna
POLAND
...INTO LATVIA, WITH ITS MANY SMALL RIVERS AND LAKES.
Poznan
Lubin
LOOKING OUT OVER THE GULF OF RIGA, I'M AMAZED TO BE SEEING ALL OF THESE PLACES I ONCE ONLY READ ABOUT.
I FEEL LIKE A GREAT CONQUEROR.

SEPTEMBER 8TH. LENINGRAD. 59° NORTH LATITUDE. THAT'S PRACTICALLY THE ARCTIC CIRCLE. IT'S ALREADY SO COLD, AND THIS IS JUST AUTUMN. THE SUN WILL BARELY RISE COME WINTER.
SUDDENLY, THE SMART UNIFORM THAT FELT SO COMFORTABLE DURING A SUMMER MARCH SEEMS VERY THIN INDEED. I REQUESTED WINTER GEAR.
HA! WE'LL BE HOME LONG BEFORE THEN!
OCTOBER 16TH. LENINGRAD IS WELL DEFENDED BY ITS SOLDIERS AND ITS CITIZENS. WE ARE UNABLE TO OVERRUN IT, AND SO THE ORDER COMES TO LAY SIEGE. WE WILL CUT THE CITY OFF FROM THE WORLD AND STARVE IT INTO SURRENDER.
WE HAVE OUR OWN TROUBLES THOUGH. THE RAINS ARE CONSTANT, AND OUR SUPPLY TRUCKS CAN'T GET THROUGH ROADS THAT HAVE BECOME MUD PITS. THAT MEANS NO WOOL CLOTHES, FRESH FOOD OR MEDICINE. WHAT WILL HAPPEN TO US WHEN OUR SUPPLIES RUN OUT?

NOVEMBER 1ST. NEW ORDERS. WE'RE ON THE MOVE AGAIN, SOUTH, TOWARDS MOSCOW. SOUTH DOESN'T MEAN WARMER, AS WE LEAVE THE TEMPERING INFLUENCE OF THE BALTIC SEA.
THE RAINS HAVE TURNED TO SNOW. SUPPLIES ARE STILL SCARCE, AND THE FLEEING SOVIETS HAVE BURNT THEIR HOMES, THEIR BARNS, THEIR CROPS AND EVEN DESTROYED THEIR LIVESTOCK RATHER THAN LET US HAVE ANYTHING OF USE.
NOVEMBER 15TH. WE CAN SEE MOSCOW IN THE DISTANCE, BUT WE HAVE MET STRONG RESIST-ANCE. THE SOVIETS FIGHT HARD. STILL, WE MUST TAKE THIS CITY.

DECEMBER 8TH. I CAN'T DESCRIBE HOW COLD IT IS. WE CAN BARELY MOVE FOR FEAR OF SNAPPING LIKE ICICLES. OUR MIGHTY TANKS SPUTTER AND STOP AND OUR GUNS FREEZE UP; NOTHING WORKS IN THIS TERRIBLE COLD.
NOTHING WORKS OF OURS, ANYWAY. THE SOVIET SOLDIERS, OUTFITTED WITH WARM, HEAVY COATS, DON'T SEEM TO EVEN NOTICE THE WEATHER, AND THEY KNOW HOW TO PROTECT AND MAINTAIN THEIR EQUIPMENT IN THESE NIGHTMARE CONDITIONS.
IN THEIR WHITE CAMOUFLAGE, THEY DISAPPEAR INTO THE SNOW LIKE GHOSTS. WE CAN BARELY RESIST THEIR ATTACKS. WE ARE FALLING TO SICKNESS, EXHAUSTION, AND FROSTBITE.
WE CAN'T WIN THIS.

JANUARY 7TH, 1942. FINALLY, THE ORDER COMES: WE WITHDRAW 250 KM FROM MOSCOW EVEN AS SOVIET REINFORCEMENTS ARRIVE FROM SIBERIA.

READING BACK THROUGH THIS JOURNAL, I LAUGH BITTERLY AT MYSELF. I THOUGHT THIS WAS A NATURE HIKE? THAT I WAS A CONQUEROR? PART OF AN IRRESISTIBLE ARMY?

WE HAVE BEEN BROKEN BY THE RUSSIAN WINTER. THERE IS NOTHING LEFT TO DO, HERE AMIDST SO MUCH BLOOD AND SNOW, BUT DREAM OF SEEING HOME AGAIN.

LIEUTENANT MAXWELL KNOPF, WEHRMACHT ARMY

Unit 3

Russia

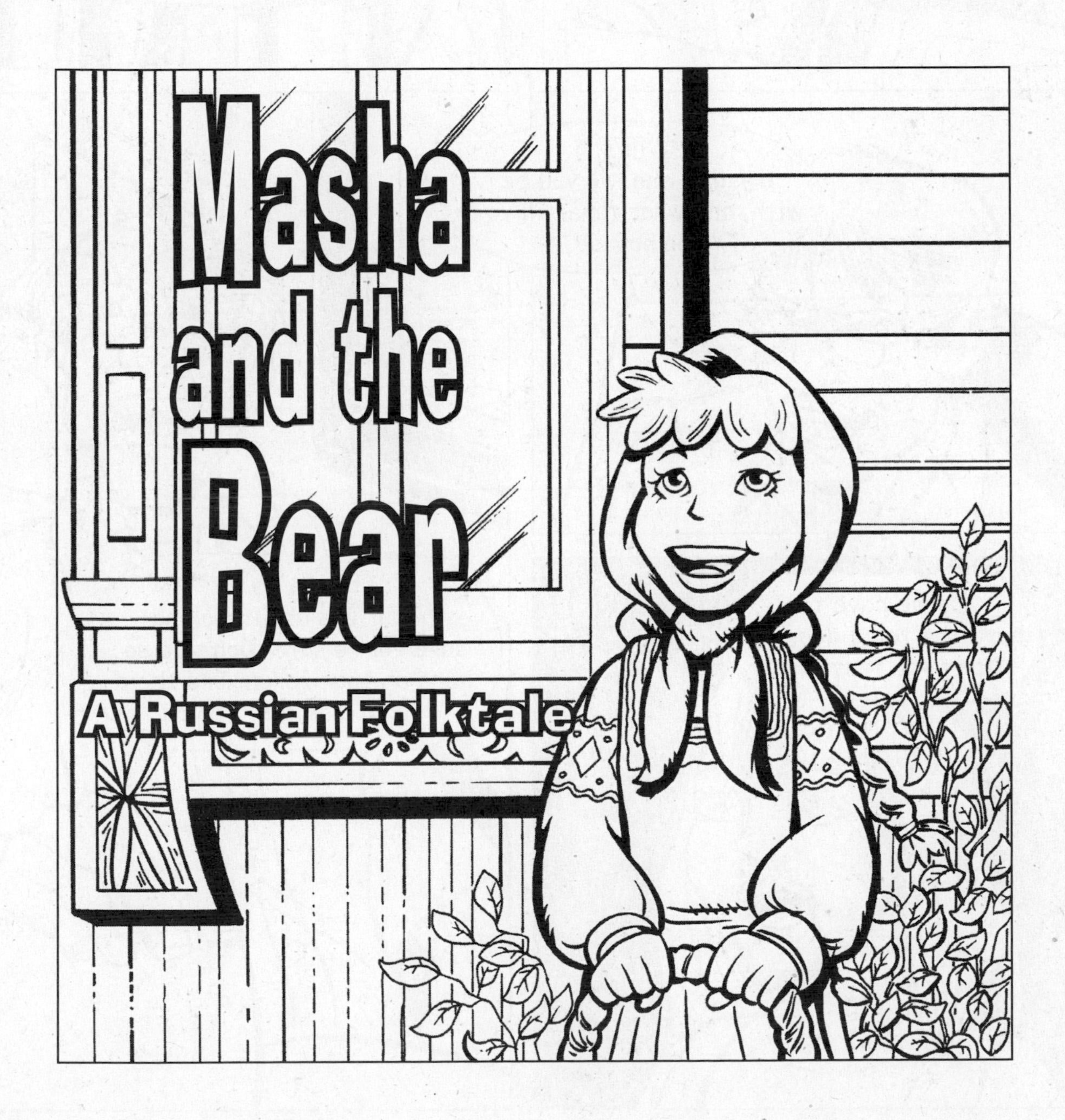

I'm going to the mall tonight, okay?

Whoa, hold on. Is your homework done?
Yes.

Who's driving, who will you be with, and what time will you be home?

Do we have to have the third degree? Don't you trust me?

It's not about trust. It's about making sure you're safe. Don't be so hard on your folks. It's their job to look out for you.

Come on over here and sit down by me.
Some things never change. I worried about your dad when he was your age, and my parents worried about me.
Let me tell you an old story my Papa told me in Russia when I wanted to go out with my friends.

There once was a clever girl named Masha, about your age, or maybe a little younger, who lived with her grandparents. One day, she wanted to pick berries and mush-rooms in the forest with her friends. Her grandparents were worried she might get lost in the darkness of the forest, but she promised to stay with her friends and be very careful.
I'll be careful, I promise.

Aleksi? Anna?
Indeed, as Masha and her friends went from bush to bush into the forest, they managed to get separated and out of earshot of each other.

Masha wandered alone until she came to a little house in a clearing.
Hello? Anyone home?
I wonder who lives here.
There was no one home, so she went inside.
Aha!
An uninvited guest!
Unfortunately, Masha had stumbled into the home of a ferocious bear.
You will stay with me, meek as a mouse. You will cook my breakfast and my dinner, and be my servant, faithful and true.

Masha was afraid and very sad, but because she could do nothing else, she stayed with the bear and kept house for him.

You must never go out without me, or I will catch you and eat you up!

bear, went to the tree top, Masha
As the bear went to the porch to make sure the weather was fine, Masha climbed into the basket.

True to his word, and thinking Masha was already watching him, the bear set off for the village with the basket strapped to his back.

If I don't rest my bones, I think I will collapse, so I will sit on a stump and eat a pie.

Don't sit on the stump and don't eat my pie. Take it to Grandma and Grandpa.
Dear me, what sharp eyes Masha has. She sees everything!

What a clever girl Masha is, watching me from atop the tree.
Several times the bear was sorely tempted to stop to rest and eat one of Masha's pies, but each time Masha called out from the basket, "I see you, I see you." Each time he picked up the basket and went on even faster toward the village.

Masha's grandparents were overjoyed to find Masha delivered safely back to them by the very bear who had held her captive for so long. They readily agreed that Masha was a very clever girl .

Unit 4

North Africa, Southwest Asia, and Central Asia

BENI HASAN, EGYPT
STARMORE
So here we are, girls, at the Starmore Center for Anthropology!
Hope you girls are up for some serious learning!

I don't know, Grandpa. Aren't most museums pretty BORING?
Oh, not this one, Yasmeen. And that's a promise.
Hey, who's that?

Oh, that's Lady Thomasina Starmore.
She founded this museum.

I LOVE her cat!
She looks like a real explorer!

Yes, she was certainly UNIQUE.
Lady Starmore studied the ancient Egyptian goddess Bast.

She named her cat Bubastis, after the city where Bast's biggest shrine was located.
Bubastis's lovely turquoise-colored pendant is made of the special Egyptian earthenware, faience.

Yasmeen, that cat in the painting- its eyes - they're FOLLOWING us!
Oh, they are NOT!
Well, it looked like they were!

A HALF AN HOUR LATER...
So girls, like I was saying, the Egyptians first took cats into their homes 3,000 years ago.

Cats came in handy for protecting the royal granaries from rats and mice.
The Egyptians also respected cats for fighting off snakes, especially cobras.
Whoa!

The goddess Bast represented many things over the long course of the Egyptian dynasties...
motherhood, the good things the sun brings us, and protection.

As you can see, artists showed Bast as having the body of a woman and the head of a cat.
Because of their connection to Bast, cats were treated as divine.

Let's say there was a fire in an ancient Egyptian home.
The people who lived there made saving their cats the priority, even above saving family or possessions.

Really?!
Yes, cats were very important to the ancient Egyptians.
So important that they wanted cats with them in the AFTER-LIFE!

The Egyptians mummified people and animals so they would survive intact in the afterlife.

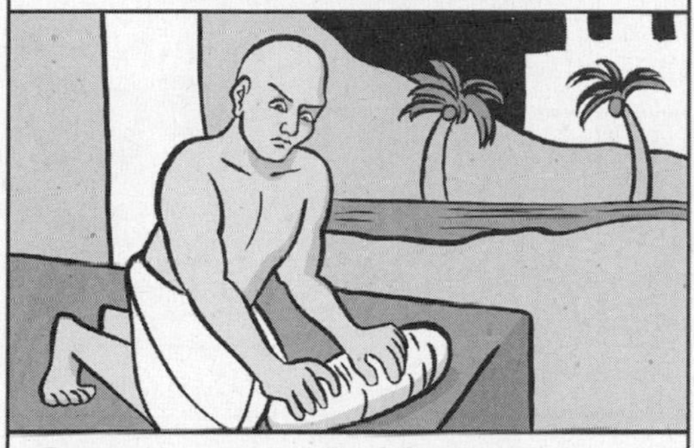

Even the cat of a farmer or a herdsman would be mummified.

The ancient Egyptians used many preservatives for making mummies.

One of them was a special salt called natron. Myrrh and other natural substances were also used.

Once preserved, the mummy was wrapped with strips of linen.

If it was a special cat, charms might be placed in the linen, and a face would be drawn on the outside of the wrapped body.

It was hard to find wood in ancient Egypt, so only the wealthiest could afford coffins for their cat mummies.
But everyone made sure to bury their mummified cats in special cemeteries that were only for felines.

Well, girls, your grandmother is waiting for us in the cafe. Want to take a break for some tea and snacks?
Yeah!

LATER THAT DAY...
That painting of Lady Starmore really creeped me out.
More than the mummies? That's hard to believe.

Oh my!
What's this?

Mrrrr?
It's the cat from that painting!

Agh! It's a ghost!
GIRLS!
The cat with the moving eyes!

You and your pranks!
Why couldn't you just tell them that we adopted one of Bubastis's kittens and inherited that collar?

Come back! Amira! Yasmeen!

Oh, don't mind them, little one.
Now tell me - where have you been hiding all day?
E. Lindner

Unit 4

North Africa, Southwest Asia, and Central Asia

Nebuchadnezzar's

Babylon

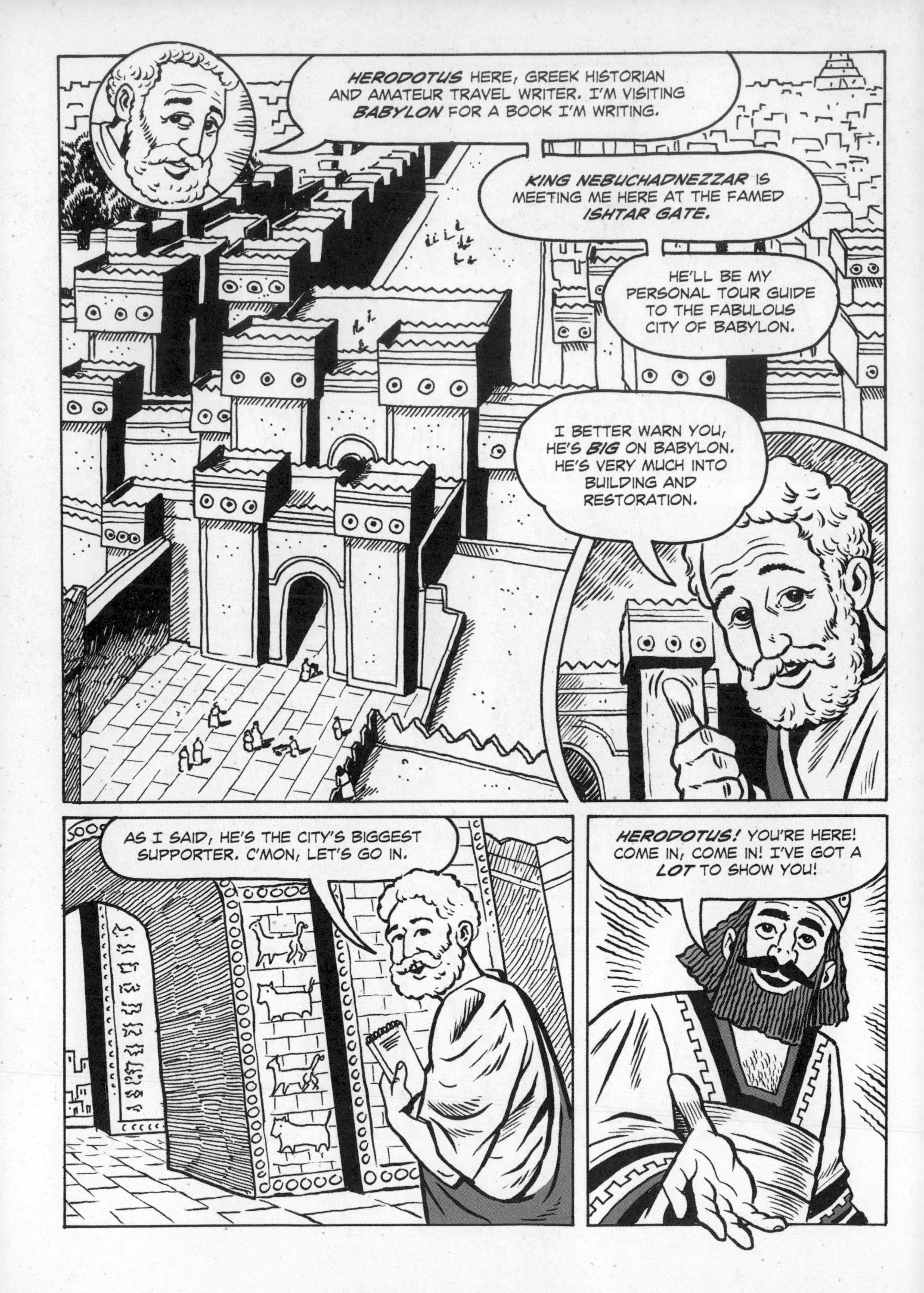
HERODOTUS HERE, GREEK HISTORIAN AND AMATEUR TRAVEL WRITER. I'M VISITING BABYLON FOR A BOOK I'M WRITING.
KING NEBUCHADNEZZAR IS MEETING ME HERE AT THE FAMED ISHTAR GATE.
HE'LL BE MY PERSONAL TOUR GUIDE TO THE FABULOUS CITY OF BABYLON.
I BETTER WARN YOU, HE'S BIG ON BABYLON. HE'S VERY MUCH INTO BUILDING AND RESTORATION.
AS I SAID, HE'S THE CITY'S BIGGEST SUPPORTER. C'MON, LET'S GO IN.
HERODOTUS! YOU'RE HERE! COME IN, COME IN! I'VE GOT A LOT TO SHOW YOU!

KING NEBUCHADNEZZAR, IT'S SO NICE TO FINALLY MEET YOU.
OH PUH-LEEZE! JUST CALL ME *KING NEB.*

THIS IS QUITE AN ENTRANCE.
IT *IS* IMPRESSIVE, ISN'T IT? WE CALL THIS ROADWAY, *PROCESSIONAL WAY.*

IT'S 80 FEET WIDE AND MADE OF LIMESTONE AND PINK MARBLE. IT'S *GREAT* FOR PARADES!

BUT I'M REALLY CRAZY ABOUT THESE BAS RELIEF LIONS, BULLS, AND DRAGONS BUILT INTO THE GATE WALLS. ALL THOSE ENAMELED TILES IN BRIGHT COLORS. AREN'T THEY *WILD?*
OH YES, I SEE THAT THE LIONS ARE OUTSIDE THE GATE WALLS AND THE BULLS AND DRAGONS ARE WITHIN THE GATE.

OF COURSE, I MUST SHOW OFF THE HANGING GARDENS BEFORE WE GO MUCH FURTHER. I'M SURE YOU'VE HEARD THAT MY GARDENS JUST MADE THE SEVEN WONDERS OF THE ANCIENT WORLD LIST!
THE STORY GOES THAT NEBUCHADNEZZAR BUILT THE HANGING GARDENS TO PLEASE HIS YOUNG WIFE WHO CAME FROM A MOUNTAINOUS AND GREEN PART OF PERSIA. BECAUSE BABYLON WAS SURROUNDED BY DESERT, SHE GREW HOMESICK FOR THE LUSH GREENERY OF HER HOMELAND.
I HAD TO BUILD THEM NEAR THE EUPHRATES RIVER TO IRRIGATE ALL THE PLANTS, SEE HOW THEY RISE UP TO ALMOST 75 FEET?
THEY'RE BEAUTIFUL. HOW DID YOU DO THIS?
AFTER I BUILT THE BASIC STRUCTURE, THE WHOLE BUILDING WAS WATERPROOFED WITH BITUMEN, BAKED BRICK, AND LEAD-MAINLY TO KEEP THE LOWER VAULTS DRY. FINALLY, I COVERED THE TERRACES WITH ENOUGH EARTH TO SUPPORT EVEN THE LARGEST TREES. SPECIALLY BUILT MACHINERY CARRIES WATER UP TO THE HIGHEST TERRACES.

THESE GARDENS ARE TRULY REMARKABLE. THEY ARE AN OASIS IN THE MIDDLE OF THE DESERT. YOUR WIFE MUST BE VERY HAPPY.
AMYTIS ***LOVES*** THIS PLACE! SHE SPENDS HOURS UP HERE, SITTING IN THE SHADE OR TENDING THE FLOWERS.
OH, THERE'S THE FABLED TOWER!
AH, YES. WE JUST RENOVATED THE TOWER, YOU KNOW. THAT'S OUR NEXT STOP. WAIT 'TIL YOU SEE THE VIEW FROM THE SUMMIT. IT'LL KNOCK YOUR SANDALS OFF!
DID YOU KNOW THAT A TOWER HAS EXISTED ON THE SITE OF OUR NEW TOWER FOR CENTURIES!
LIKE THEY SAY IN THE REAL ESTATE BIZ, "LOCATION, LOCATION, LOCATION!"

THE SIDES OF THE BASE ARE EACH 300 FEET, AND COVERED IN A DECORATIVE CASING OF BURNT BRICK. THE MAIN STAIRWAY IS 30 FEET WIDE.
THAT'S QUITE A CLIMB TO THE TOP!

BUT IT'S WELL WORTH THE EFFORT, MY FRIEND! C'MON, LET'S GET STARTED.

AT THE HALFWAY POINT...
HOW HIGH *IS* THIS TOWER, ANYHOW?
300 FEET, BUT LOOK AT THE VIEW JUST AT THIS HEIGHT. SPECTACULAR, HUH?

FINALLY, THEY REACH THE SMALL TEMPLE AT THE SUMMIT.
WAS I RIGHT, OR WHAT? QUITE A VIEW!
WHEW! ***MAGNIFICENT!***

I DIDN'T REALIZE THE CITY WAS SO BIG!
OH YES! IT'S LAID OUT IN A VERY LARGE SQUARE–14 MILES ON EACH SIDE. I'M VERY CONCERNED ABOUT SECURITY FOR MY PEOPLE, SO I BUILT A DOUBLE-WALLED SYSTEM SURROUNDING THE CITY.
THE MAIN WALL IS 56 MILES LONG AND VERY HIGH. THEN THERE'S A SECOND WALL BEHIND THAT ONE. IT'S A GREAT DETERRENT TO INVADING ARMIES, AS IS THE WIDE, DEEP MOAT AT THE BASE OF THE OUTER WALL.
THE WALL ALSO HAS SEVERAL TALL TOWERS FOR GREATER VISIBILITY. THERE ARE STURDY MASSIVE GATES LEADING INTO THE CITY, SO WE ARE VERY SECURE BEHIND OUR WALLS.
I'VE NEVER LAID EYES ON SUCH SIGHTS!
YOU'LL GIVE US 5 STARS, THEN?
I'D GIVE BABYLON 10 STARS, KING NEBUCHADNEZZAR!
IT'S HARD TO BELIEVE, BUT CENTURIES LATER ALL THAT WOULD REMAIN OF BABYLON'S MAGNIFICENCE WOULD BE SOME MOUNDS OF EARTH IN THE PERSIAN DESERT.

UNIT 5

Africa South of the Sahara

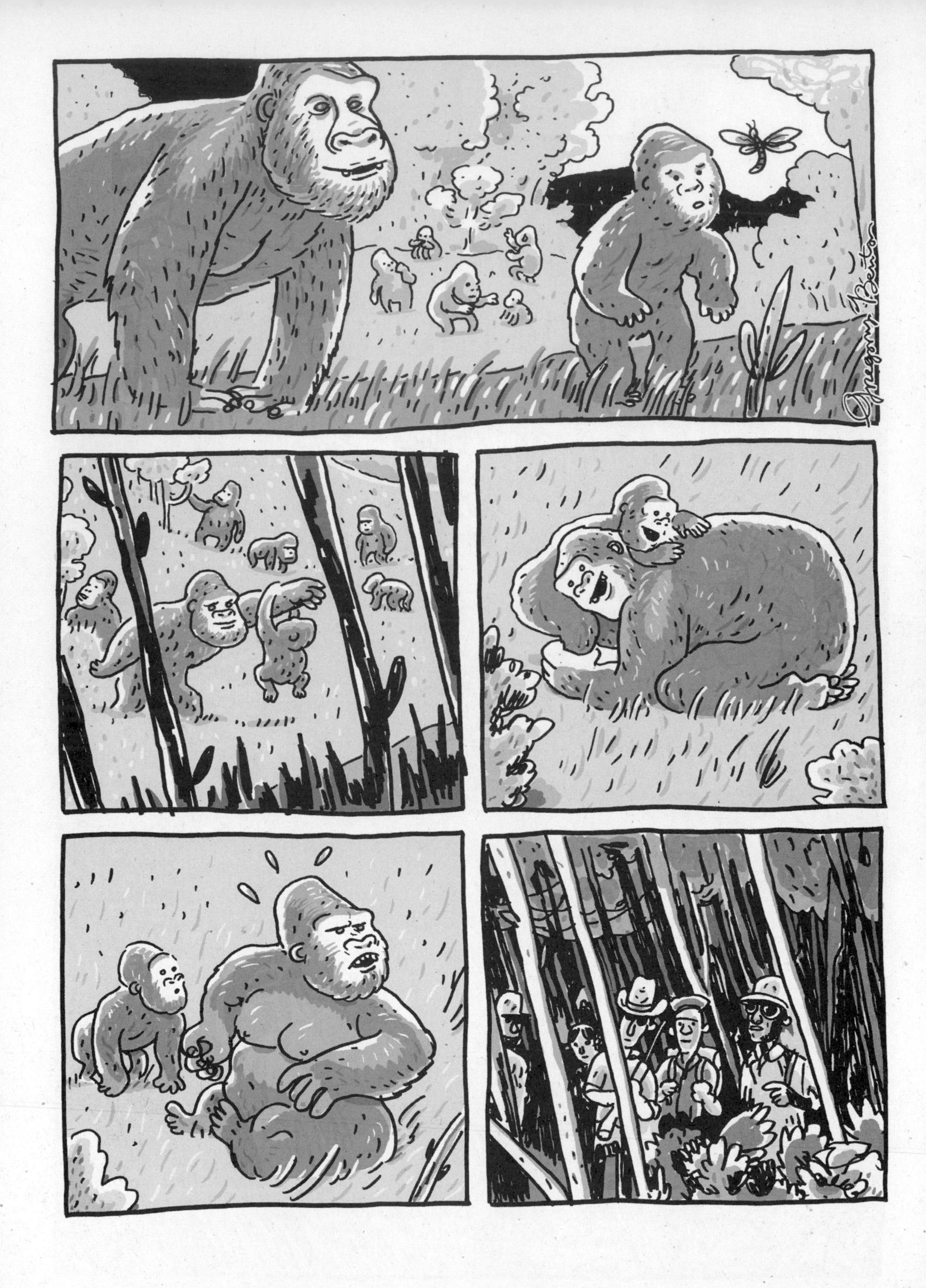
Gregory Benton

THERE THEY ARE–
RARE MOUNTAIN GORILLAS.
GUIDE

WHILE THEIR NUMBERS DWINDLED FOR MANY YEARS, THEIR POPULATION IS FINALLY STARTING TO GROW.

THESE GORILLAS ARE STILL ENDANGERED, AND HUMAN ACTIVITIES GREATLY THREATEN THEIR SURVIVAL.

IF THEY ARE TO SURVIVE, PEOPLE MUST BE THE ONES TO HELP.
end

UNIT 5

Africa South of the Sahara

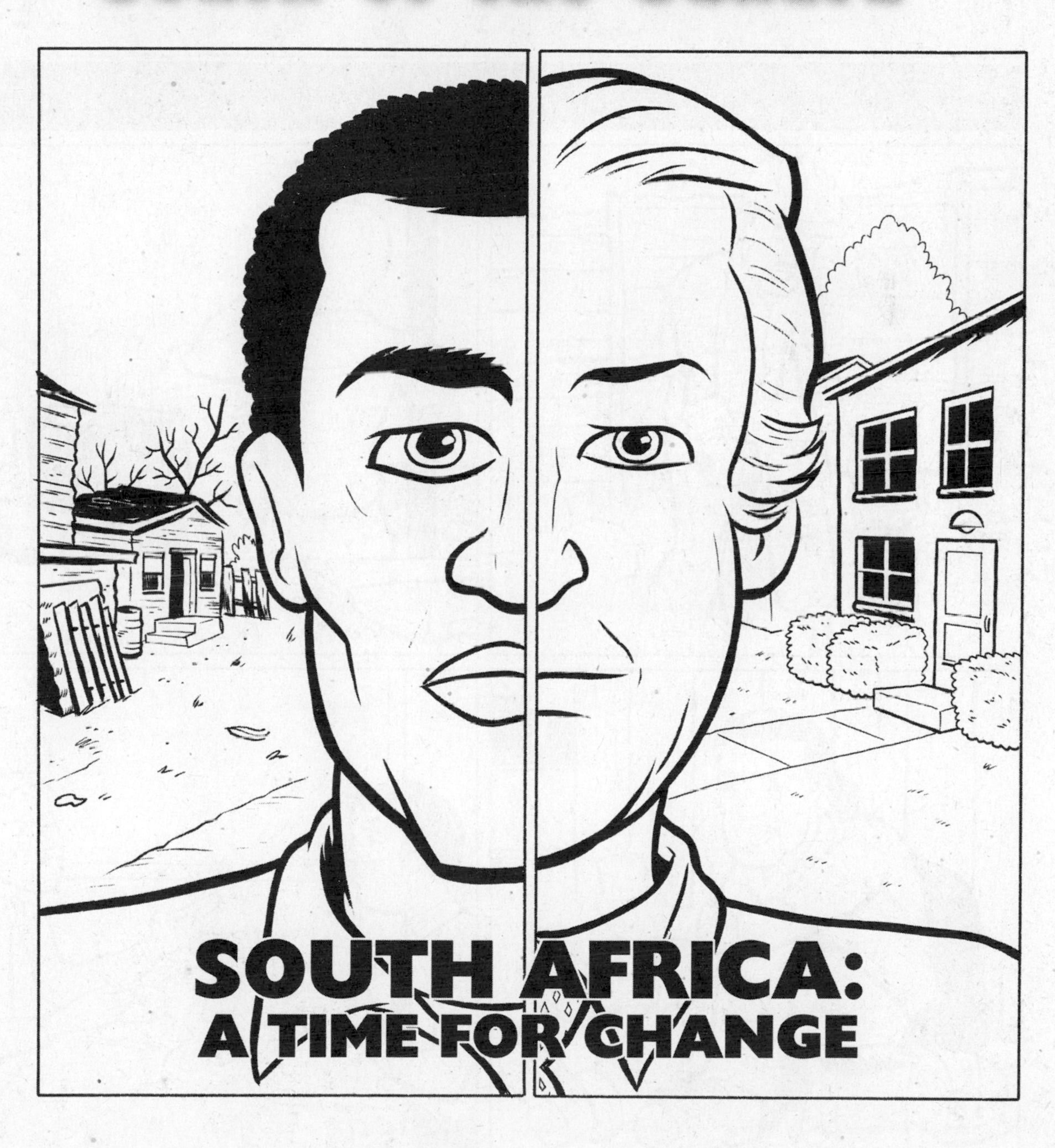

CAPE TOWN, SOUTH AFRICA. 1994.
DVORAK CONSTRUCTION

NEWS
ELECTIONS
TOMORRO
VOTE
ELECTION
DAY

VOTE!
APRIL
POLLING STATION
PLACE BALLOTS HERE

Unit 6

South Asia

SHERPAS TO THE RESCUE

MORNING, DAY 1:
DUDE, I CAN'T BELIEVE IT. WE ARE IN NEPAL AND GOING TO CLIMB MOUNT EVEREST!
I KNOW, DUDE. THE HIMALAYAS ARE TOTALLY AWESOME!
THIS WILL BE THE ADVENTURE OF A LIFETIME!

'MORNING, ANG. WHO IS IT THIS TIME?
AMERICANS.

OH, RIGHT. "ADVENTURE OF A LIFETIME"?
OF COURSE.

HA HA HA HA
HA HA
HA

WELL, LET'S GET TO WORK.
ANOTHER DAY, ANOTHER DOLLAR.

AFTERNOON:
DUDE, THIS IS SO EXTREME!
WE'RE GOING TO BE HEROES BACK HOME!

I BET WE GET OUR PICTURES IN THE PAPER!
HOW MANY PEOPLE CAN SAY THEY DID THIS?
...SO I SAID, "I'VE DONE THIS DOZENS OF TIMES. TRUST ME. THE ROPE'LL HOLD."

HEH HEH... THEY ALWAYS LIKE TO THINK THEY'RE THE FIRST TO EVER DO THIS, DON'T THEY?

EVENING:
THAT'S ALL FOR TODAY. WE CAMP HERE.
AWESOME! I'M BEAT.

THIS ROCKS, BRO. THERE IS NOTHING LIKE PUSHING YOUR BODY TO THE ABSOLUTE LIMIT, THEN WHEN YOU CAN'T GO ANY FURTHER, TAKING A WELL-DESERVED REST.
YEAH.

WE TOTALLY EARNED THIS. TODAY, WE WORKED AS HARD AS HUMANLY POSSIBLE.

OKAY, I SET UP CAMP, AND ANG WILL HAVE DINNER READY SHORTLY.

THAT NIGHT:
GOT EVERYTHING CLEANED UP?
YEAH. TIME FOR BED.

SEE YOU IN FOUR HOURS THEN, ANG.
GOOD NIGHT, TASHI.

DAY 34:
THE SUMMIT OF MOUNT EVEREST, THE TALLEST MOUNTAIN IN THE WORLD!
WE'RE MORE THAN 29,000 FEET UP! WE RULE!
YEAH. THINK ABOUT IT, MAN.

WE FACED NATURE'S GREATEST CHALLENGES ARMED ONLY WITH OUR WITS AND OUR BODIES, TRAINED TO PEAK PHYSICAL PERFECTION.

OH BROTHER.

DUDE! TAKE MY PICTURE!
OKAY, BUT YOU HAVE TO TAKE MINE, TOO!

THOSE TWO ARE NEVER GOING TO MAKE IT BACK DOWN THE MOUNTAIN ON THEIR OWN.
NOT A CHANCE.

DAY 35:
HELLO?
CAN YOU HEAR ME?

WHAT? WHERE AM I, DUDE?
MEDICAL HELICOPTER. YOU GUYS ARE SUFFERING FROM FROSTBITE AND SEVERE EXHAUSTION.
HOW'D WE GET HERE?

YOUR GUIDES CARRIED YOU BOTH DOWN THE MOUNTAIN ON STRETCHERS.
LOOKED LIKE THEY MADE A LITTLE RACE OUT OF IT.

WE SURE CALLED THAT ONE.
AS USUAL.

A HELICOPTER!
THIS STINKS!
WHAT IS YOUR PROBLEM? THOSE SHERPAS SAVED YOUR LIVES!

YEAH, BUT UP TILL THIS, WE WERE DOING IT ALL ON OUR OWN!
!

Unit 7

East Asia and Southeast Asia

IN 1989 ABOUT 100,000 STUDENTS AND WORKERS GATHERED IN BEIJING, CHINA, IN TIANANMEN SQUARE. THE PRO-DEMOCRACY MOVEMENT PLANNED A PEACEFUL POLITICAL PROTEST.
THE PROTESTORS DEMANDED FREE SPEECH AND A FREE PRESS.
THEY WANTED THE AGING LEADERS OF CHINA'S COMMUNIST PARTY TO RESIGN.

THE PROTESTORS KNEW THE NEWS NETWORKS FROM THE ENTIRE WORLD WOULD BE IN CHINA TO COVER A VISIT FROM THE SOVIET UNION'S LEADER, MIKHAIL GORBACHEV. THEY USED THIS INTERNATIONAL COVERAGE TO VOICE THEIR OPINIONS ABOUT POLITICAL CORRUPTION IN THE COMMUNIST GOVERNMENT IN CHINA.
ART STUDENTS BUILT A LARGE STATUE OUT OF PAPIER-MÂCHÉ MODELED AFTER THE UNITED STATES' STATUE OF LIBERTY. THE STUDENTS CALLED IT "THE GODDESS OF DEMOCRACY" AND RALLIED AROUND THIS SYMBOL OF FREEDOM.

THE PROTESTORS SANG SONGS AND MARCHED AROUND THE SQUARE. STUDENTS FROM THE COLLEGES IN THE AREA BOYCOTTED THEIR SCHEDULED CLASSES.
THE NEWS COVERAGE OF THE PROTESTS AIRED NONSTOP ON TELEVISION NETWORKS AROUND THE WORLD.
OUT OF DESPERATION, HUNDREDS OF STUDENTS WAGED HUNGER STRIKES.
ALTHOUGH THE PROTEST REMAINED PEACEFUL FOR WEEKS...
A VIOLENT CLASH WAS INEVITABLE.

COMMUNIST PARTY LEADERS DECIDED TO USE MILITARY FORCE TO TAKE CONTROL OF THE SQUARE. SOLDIERS AND TANKS FROM THE PEOPLE'S LIBERATION ARMY CLASHED WITH UNARMED CHINESE STUDENTS AND WORKERS IN THE STREETS OF BEIJING. THE SOLDIERS FIRED AT RANDOM IN TO THE CROWD. IN THE CHAOS, HUNDREDS WERE KILLED AND THOUSANDS WOUNDED.

AFTER THE VIOLENT SHOWDOWN, THE GOVERNMENT SUPPRESSED THE REMAINING DEMONSTRATORS WITH THREATS AND ARRESTS. IT WAS SO EFFECTIVE THAT STUDENTS AND PROTESTORS STOPPED TALKING TO THE MEDIA ENTIRELY.
A NEWS "BLACK OUT" WAS IMPOSED, AND THE FOREIGN PRESS CAMERAS WERE LITERALLY UNPLUGGED.
THE CHINESE PEOPLE WHO SAW THE DEMONSTRATIONS ON TV NEVER SAW THE TRUE EVENTS UNFOLDING. INSTEAD, THEY RECEIVED THE "OFFICIAL" VERSION, CREATED BY THE GOVERNMENT-RUN NEWS STATION.
THE CHINESE GOVERNMENT TRIED TO ERASE THE MEMORY OF THE TIANANMEN SQUARE PROTESTS, BUT THE STRUGGLE COULD NOT BE FORGOTTEN.

Unit 7

East Asia and Southeast Asia

In the mid-1200s the powerful Mongol leader, Kublai Khan, sent representatives to Japan, demanding that the Japanese pay tribute to the Mongol empire. The Japanese refused, and Kublai Khan drew up plans for invading Japan's islands. In 1274, a combined force of more than 20,000 Mongol and Koryo soldiers aboard 900 ships invaded Japan. The Mongol forces quickly captured several smaller islands and then landed at Kyushu. Although the samurai fought valiantly, they were no match for the Mongols' tactics and weapons. The Mongols used weapons never before seen by the Japanese, such as tetsuhau, exploding shells made of a ceramic material and filled with gunpowder.
BOOM
The samurai were forced inland. The Mongols, however, perhaps concerned about Japanese reinforcements and worsening weather, retreated to their ships. The strengthening wind shifted, wrecking a number of Mongol ships. The remaining Mongol forces departed for home.

In 1281, Kublai Khan ordered another invasion of Japan. On the islands, the samurai gathered their armies and awaited the attack. One portion of the Mongol force, some 900 ships, attacked first and again captured several small islands.
At Kyushu, however, the Japanese had built a long defensive wall along the coastline. The first wave of invaders was forced to turn back and await a much larger second force.

How many?

The second Mongol force appears to have 3,500 ships!

3,500?! That is more than three times the size of their first attack!

With that large of a fleet, they must have at least 100,000 men.

What shall we do, sir?

What do you think
we should do?
Eh?

We have no choice but to defend ourselves, sir. Perhaps we can continue to strike at the enemy's individual ships.
Yes. We had some success attacking their ships at anchor.

Our soldiers and samurai have boarded and destroyed their ships by night. Their great warships were forced to tie together for safety. In the close quarters, many of their soldiers also died from disease.

You are right. We have no choice but to defend ourselves...

but I fear the fate of our country is no longer in our hands.

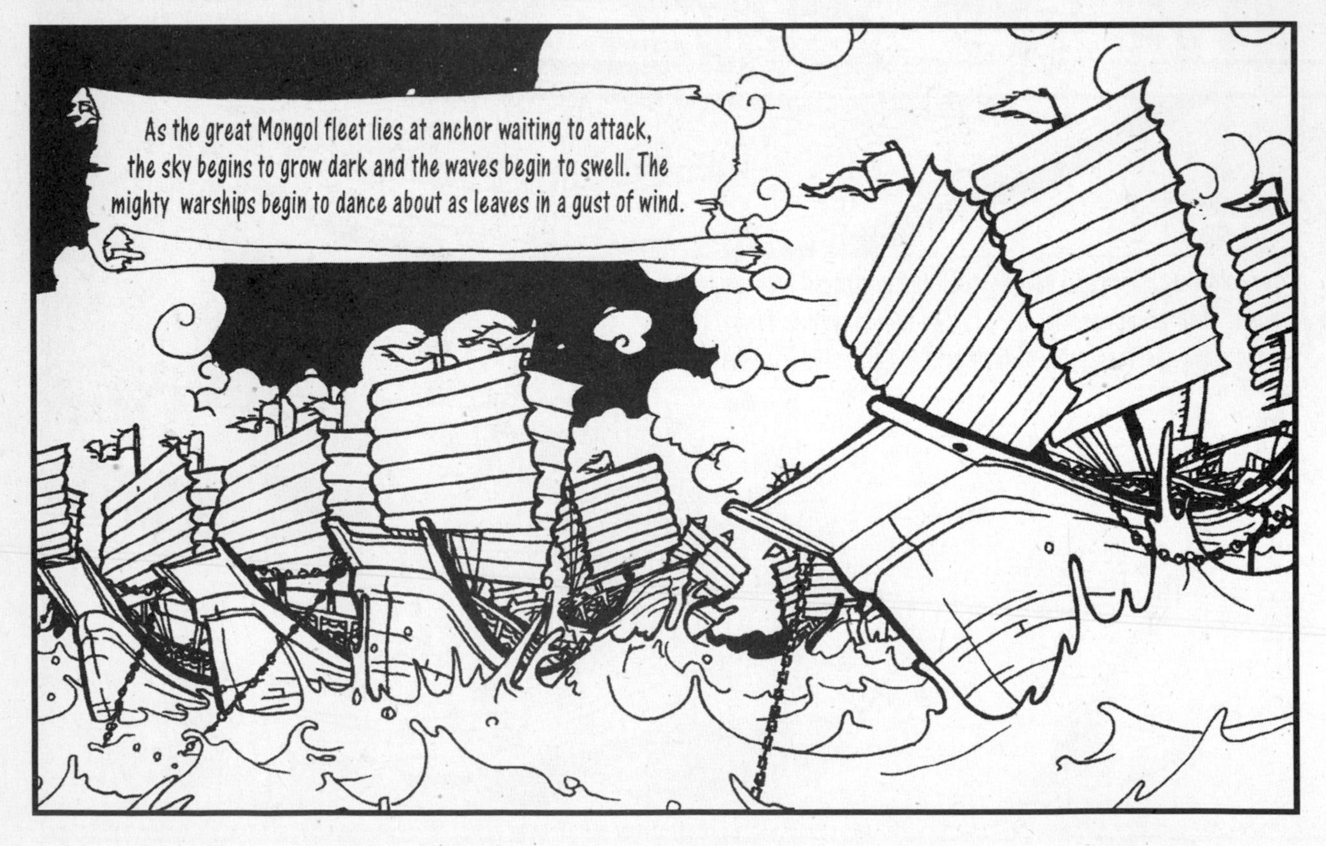
As the great Mongol fleet lies at anchor waiting to attack, the sky begins to grow dark and the waves begin to swell. The mighty warships begin to dance about as leaves in a gust of wind.

Sir! Sir!

Sir, the typhoon has struck the enemy fleet! Many of their ships were destroyed! The Mongols have retreated! We are saved.

It is truly a divine wind, a...kamikaze.

Unit 8

Australia, Oceania, and Antarctica

Good news, class! I've arranged a very special opportunity for all of you this summer.
HANALEI MIDDLE SCHOOL
You are all invited to spend the summer on a small island in the South Pacific. You'll have to use the knowledge you've learned in class to survive.
UNIT 5: PACIFIC ISLANDS GEOGRAPHY
Mr. Santiago's class set sail for a remote island in late June.
NOW what?
I'm hungry!
I'm scared!
Where do we sleep?
I can't believe Mr. S. just left us here!
Is the boat really gone?
WE'RE ALL ALONE IN THE MIDDLE OF NOWHERE!!!
We'd better figure out a place to sleep tonight.
And what we're going to EAT!
Maybe we can make a shelter out of palm leaves.

Hey, Jim, what do you think we'll find to eat here?
I'm not sure, Mari. We should be able to find bananas on these islands.
Didn't anyone bring matches?
This takes *forever!*
HEY! *It sparked! QUICK!* Someone get the kindling!
POP!
It's pitch black out here.
What's that noise?!
Ssshh! Something's out there!
Welcome! Your teacher told us you were coming.
We are here to watch over you and make sure you're safe.
Have you had anything to eat?
We made hot banana soup with coconut milk!
Yum! That sounds like dessert! My name is Pomare. Tomorrow I can show you how to catch a fish. Steam it with banana leaves and it'll be the best fish you've ever tasted!
The next morning
KOO-KEE! KOO-KEE!
ARR-ARR-OOO!
CHIRRUP CHIRRUP
No need for alarm clocks here!
HEY!! Sara! Kiko! *LOOK!!*
I caught a crab!
POK POK POK

Ah! I see Hector has already started fishing.
Hi, Pomare! We saw a movie in our class about fishing with spears!
Good work! Now - can you catch twenty more to feed the group by this evening?
Uh...he was kind of tricky to catch. Maybe everyone should get their own.
SKRITCH SKRITCH
We can't all spend the whole day fishing. Who's going to build a sturdier shelter?
We need to gather other stuff to eat, too.
Okay, okay! I'll spend all day catching twenty crabs. Sheesh!
Ha ha! Don't worry, I'll help you. I'll show you how to make a net.
Cool!
Can I help?
Of course, Kiko. We will need more than one fishing expert.
We're using ready-made coir rope, made from coconut fiber. Traditionally, we would first dry the fibers, and spin them together to make rope.
I've got more wood for you to chop!
ARGH! Don't we have enough yet?
Here are our taro and sweet potato plants.
Isn't volcanic soil good for crops?
Yes! Very good, Martin!

Summer passes quickly in "Santiago Village"...
SANTIAGO VILLAGE
We should have a contest to elect a king or queen!
Yeah! A swimming race!
Or a coco palm climbing race!
How 'bout a fishing competition?
Let's make it a triathalon!
Okay! You guys know the rules: swim out to your outrigger canoe, catch a fish, paddle to the shore, put your fish in the bucket, climb a coco tree, and bring five coconuts to the finish line!
GO!
Keep an eye on your boat!
Looking good!
PANT, PANT! Made it!
C'mon, fishes! Suppertime!
I got one!
Oh no! Everyone else is on shore already!

*taufolo = baked breadfruit
*palusami = coconut cream, onions, and meat in taro leaves

*poe = a tropical fruit pudding from the South Pacific
*poi = a Hawaiian dish made from boiled taro root

*aparima = a slow, melodic dance

Unit 8

Australia, Oceania, and Antarctica

The cold weather setting in brings back so many memories. Some are of unimaginable cold and endless ice.
Others are memories of courage and bravery, of comradeship and great leadership. They are all memories of my adventure with Ernest Shackleton and the voyage of the *Endurance*.
Ernest Shackleton was a British explorer who made several trips to Antarctica. It was back in 1914 that we set sail for the southernmost continent. Other explorers had beaten Shackleton to the South Pole, but he wanted to be the first to travel across the entire continent, from shore to shore. The journey, however, went horribly wrong and nearly cost us our lives.
Shackleton sought recruits for his 1914 Imperial Trans-Antarctic Expedition by posting signs around the harbor. He needed 28 men.
MEN WANTED
FOR HAZARDOUS JOURNEY. SMALL WAGES, BITTER COLD, LONG MONTHS OF COMPLETE DARKNESS, CONSTANT DANGER, SAFE RETURN DOUBTFUL. HONOUR AND RECOGNITION IN CASE OF SUCCESS.
SIR ERNEST SHACKLETON
Well, you can't get more honest than that. I respect a man who tells the truth, and that's the kind of man I want to work for. Where do I sign up?

It was the start of an amazing journey for every man aboard the ship.
The conditions were harsh and the work was hard. But every man did his share because we all respected the man we called "The Boss."
We departed England on August 1, 1914. We arrived at the whaling station of Grytviken, South Georgia Island, on the 5th of November. It was our last glimpse of civilization for almost two years.
In early December, we sailed from South Georgia Island into the Weddell Sea. It was unusually cold, and we soon encountered large packs of ice that slowed our progress greatly.
Boy, it's getting really cold outside.
Yes, look at the ice forming.

For the next six weeks, the ship dodged and weaved between loose ice floes, even ramming through them.
We could go no further. In January 1915, the *Endurance* became trapped in the ice. We attempted in vain to chop our way out of the ice pack. It was no use. As the ice pack drifted - first to the south, then to the north - we moved helplessly along with it.
In October, the immense pressure of the ice began to crush the *Endurance*, and we were forced to abandon ship. We removed as many essential supplies as we could.
We made camp on an ice floe. "The Boss" called it "Ocean Camp." Several weeks later, we watched in horror as the ice finished its terrible work on the *Endurance*, and the ship disappeared below the surface.

After a failed attempt to reach safe ground, we made camp on another ice floe. "The Boss" called it "Patience Camp," which I thought was a clever name.
We stayed there for more than three months. By then, the ice floe had drifted out to open water.
We launched the lifeboats and set sail, hoping for safe ground.
On April 16, 1916, all three boats landed on Elephant Island - a remote, uninhabited island far from the shipping lanes. It was the first time we stood on solid ground in well over a year.
Shackleton soon had another plan. He wasn't ready to lie down and give up easily.

His plan was extremely dangerous. He was to set sail with five other men for South Georgia Island and then return with help.
On April 24, he set out in the lifeboat *James Caird*, named after one of the expedition's main patrons. It was to be a long journey.
Well chaps, we can do nothing but wait. If "The Boss" has taught us one thing, it's to keep believing. We must not give up.
He'll be back soon. I know it.
That is quite true.
After an eight day trip across open ocean, the *James Caird* made it to the coast of South Georgia Island. The whaling station was on the far side of the island though. Shackleton and his crew still had to hike through an unmapped mountain range to reach help.
SOUTH GEORGIA ISLAND
Depart Island. Dec 5, 1915
ELEPHANT ISLAND
James Caird launched. Apr 24, 1916
Launched boats. Apr 9, 1916
Endurance sank. Nov 21, 1915
Abandoned ship. Oct 27, 1915
WEDDELL SEA
Endurance trapped in ice, Jan. 10, 1915
ANTARCTICA
Looking back at what we went through since leaving South Georgia Island, we did the best we could to survive.

While we waited, we kept busy. We played football on the ice. We sang songs. We tracked and hunted seals and burned their blubber for fuel. Anything that kept our spirits up, we would do.
After finally reaching the whaling station, "The Boss" made several attempts to rescue those of us left behind. On May 23, 1916, he borrowed a ship called *Southern Sky* and tried to sail back to us on Elephant Island, but the ice was too thick. He turned back. He tried two more times in June and July, but again, he had no luck. He couldn't make another attempt until late August.
On August 30, 1916, four months after he left, Ernest Shackleton arrived on the *Yelcho* to rescue us. Our long ordeal was over, and, amazingly, every crew member had survived. It was an adventure we will never forget.